Electromechanics in Quantum Circuit Cavity

Etta E. Hicks

Acknowledgements

First of all, I'd like to thank my advisor Prof. Keith Schwab for his guidance and mentorship over the course of my study. All of this work would not be possible without his advice and continuous support. I really appreciate the freedom and resources he provided, which allow me to explore various ideas.

I have been fortunate to work with many amazing people in the course of this research. All of the works presented in this thesis were joint efforts of Junho Suh, Ari Weinstein, Emma Wollman and myself. Junho Suh and Ari Weinstein taught me all the nanofabrication and cryogenic techniques, without your guidance and experience, it won't be possible for me to conduct this research. My sincere thank also goes to our theoretical collaborators Prof. Aashish Clerk, Prof. Florian Marquardt, Andreas Kronwald, Anja Metelmann, and Matt Woolley for your great patience and persistent support.

I'd also like to thank other members in the Schwab group. Kin Chung Fong, Laura de Lorenzo, Harish Ravi, Aaron Pearlman and Steve Caton, all of you are very supportive and always willing to help. To people in the Eisenstein group, Johannes Pollanen, Chandni U, and Debaleena Nandi, thank you for all your unlimited optimism and encouragement. To people in the Painter group, Simon Groeblacher, Johannes Fink, Taofiq Paraiso, Ale Pitanti, Amir Safavi-Naeini, Jeff Hill, Richard Norte, Alex Krause, Tim Blasius, Justin Cohen, and Sean Meenehan, thank you for all your help and support, I won't forget those interesting parties and weekly soccer games.

To my friends, Hao Chu, Min-Feng Tu, Yun Mou, Jiun-Yann Yu, Chao-Yuan Yeh, Lucas Peng, Wei-Hsun Lin, Shu-Hao Liou, Evan Miyazono, Karthik Seetharam, Xiaohang Quan, and Jeff Ellis thank you for all your support and accompany, my Ph.D. journey would be colorless without you guys.

Abstract

Quantum mechanics is a very successful theory which precisely describes the behavior of microscopic particles. Although the theory provides counter-intuitive predictions, such as that an object can be at two different positions at the same time and the "spooky action at a distance" between two objects, it describes the microscopic world with unprecedental precision. These counter-intuitive quantum behaviors of microscopic particles explain phenomena across disparate scales, from the microscopic scale such as interactions between sub-atomic particles, the atomic spectrum, motion of electrons in solids, and chemical reactions, to macroscopic scale such as behavior of superfluids and superconductors, and even the evolution of stars. So far, no example of conflicts between experiment and theory is known, and the validity of the theory is so broad that one would expect that the quantum theory can be applied to the macroscopic world, for example, a macroscopic massive object such as a soccer ball can in principle be at two positions at the same time. A lot of effort has been made to observe these wierd quantum effects in macroscopic objects, and superposition of a molecular version of soccer ball, a fullerence molecule, has been observed. However, quantum phenomena of human scale objects haven't been observed due to the decoherence effects induced by the unmonitored environmental degrees of freedom, which a macroscopic object is usually coupled to.

The purpose of this dissertation is to study the quantum phenomena of a massive macroscopic object with a cavity electromechanical system, where a nanomechanical resonator couples to a high Q microwave resonator formed by superconducting circuit. Mesoscopic scale mechanical resonators have several unique properties that make it suitable for the study of macroscopic quantum phenomena: first, they consist of a macroscopic number of particles ($N \sim 10^{14}$), much more than molecules and a few order of magnitude larger than

Bose-Einstein condensate of atomic gases. Second, they are highly engineerable, and have a wide range in resonance frequency (from kHz to GHz) and possess extremely high quality factor ($Q \sim 10^8$) with meticulous design. Third, they can be macroscopic in size (a few tens of microns) while keeping the mass relatively small (~pg). All thuis together with the accessibility to cryogenic temperature (~mK) and availability of ultrasensitive detectors in mesoscopic scale puts these nanomechanical structures into the quantum regime.

In this work, we focus on the quantum effects stemming from the Heisenberg uncertainty principle. The first issue is the fundamental limitation to the precision of continuous measurement. According to the Heisenberg uncertainty principle, measuring an object would unavoidably perturbing it. Although the perturbation is very small, it would finally limit the sensitivity when the precision of the measurement is high enough. This is the subject of the first experiment in this dissertation. In this experiment, we detect this quantum backaction, which is the radiation presure from the microwave quantum fluctuation in this case, with a nanomechancial resonator. Moreover, we employ a special technique called backaction-evading measurement (or quantum non-demolition) to avoid this measurement backaction.

The second issue is the fundamental limit to the quantum state of an object. According to the Heisenberg uncertainty principle, nothing is completely at rest. Even if one cools a harmonic oscillator to the quantum ground state, it would not stay still, as there is random motion associated with the zero-point energy: the zero-point motion. This quantum fluctuation places a fundamental limit on the minimum motion of an object. One way to go beyond this limit is to squeeze the motion. By squeezing the motion, one can generate quantum squeezed states which have fluctuations below the zero-point motion in one quadrature at the expense of increasing the fluctuation in the other quadrature. These states have been generated in microscopic objects such as photons and trapped ions. In the second experiment, we employ a dissipative squeezing scheme in a cavity electromechanical system to generate a stationary quantum squeezed state of a nanomechancial resonator with more than 3 dB squeezing.

Contents

List of Figures

List of Tables

Chapter 1

Introduction

1.1 The Heisenberg uncertainty principle

The Heisenberg uncertainty principle, one of the tenets of quantum theory, was formulated by Werner Heisenberg in 1927. It states that it is impossible to measure simultaneously the canonical position q and its conjugate momentum p of an object with arbitrary precision [ ], which is

$$\Delta q \Delta p \geq \frac{\hbar}{2}. \tag{1.1}$$

As described by Heisenberg [ ], "the uncertainty principle refers to the degree of indeterminateness in the possible *present* knowledge of the *simultaneous* values of various quantities with which the quantum theory deals."

Wave properties of matter

The uncertainty relation can be deduced from the wave properties of matter. Here we follow the argument by Heisenberg in his Chicago lectures of 1930 [ ]. Consider a wave packet with spatial extent Δx made up by superposition of sinusoidal waves with wavelengths near λ_0. Then there is roughly $n = \Delta x / \lambda_0$ crests or throughs within Δx. Outside the boundary of the wave package the component waves must be canceled by interference, and therefore the set of the component waves must contain at least $n + 1$ waves that fall in the critical region. This gives

$$\frac{\Delta x}{\lambda_0 - \Delta\lambda} \geq n + \frac{1}{2}, \tag{1.2}$$

where $\Delta\lambda$ is the extent of the wavelengths of the component waves. If we expand the left hand side of Eq. (1.2) to first order in $\Delta\lambda$, we obtain

$$\frac{\Delta x \Delta\lambda}{\lambda_0^2} \geq \frac{1}{2}. \tag{1.3}$$

Together with the de Broglie relation $p = h/\lambda$, the group velocity of the wave package is given by

$$v_g = \frac{dE}{dp} = \frac{p}{m} = \frac{h}{m\lambda_0}, \tag{1.4}$$

where $E = p^2/2m$. The spreading of the wave package is characterized by the spreading of the group velocities

$$\Delta v_g = \frac{h}{m\lambda_0^2}\Delta\lambda. \tag{1.5}$$

By defining $\Delta p = m\Delta v_g$ and therefore by Eq. (1.3), we obtain the uncertainty relation

$$\Delta x \Delta p \geq \frac{\hbar}{2}. \tag{1.6}$$

In this respect, the Heisenberg uncertainty relation is the result of the intrinsic wave-like nature of the object. It specifies the limit within which the particle picture can applied or the concept of motion and path is meaningful, i.e., position and momentum of the particle are well defined simultaneously.

Heisenberg's microscope

On the other hand, the uncertainty relation can also be deduced without explicit use of the wave picture, as described by Heisenberg with his famous Heisenberg's microscope experiment [ ], which he used to give a physical interpretation of the uncertainty relation. The original version of the Heisenberg's microscope experiment considered the measure-

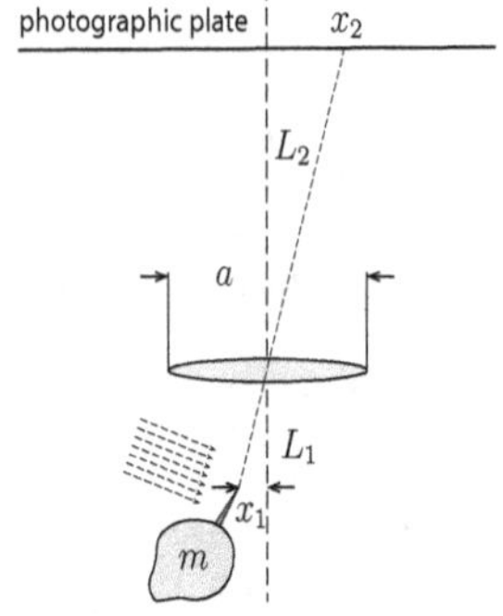

Figure 1.1: Schematic of Heisenberg's microscope experiment.

ment of the position of a microscopic particle such as electron. Here, we will describe a modified version given by Braginsky et al. [ ], which is closer to the subject discussed in this thesis. We consider measuring the position of a macroscopic object with mass m. In order to give a well defined position of the object, we attach to the object a stick with diameter less than or of the order of the wavelength of light. The position of the stick x_1 is measured by arranging it close to the focal plane of the lens. The arrangement of the lens and the photographic plate is shown in Fig. (1.1), where a is the diameter of the aperture and L_1 is the focal length of the lens. To measure the position of the stick, we send a stream of photons with wavelength λ from the side and wait for an *individual* photon to be scattered by the stick, pass through the lens's aperture, impinge on the photographic plate, collapse, and produce a small seed of silver. The transverse position x_2 of the silver seed, relative to the lens's optical axis, can be determined to an accuracy much better than an optical wavelength. Then the transverse position of the stick relative to the optical axis of the lens can be inferred with the relation $x_1 = -x_2 L_1/L_2$. The imprecision of the position measurement is limited by the Abbe diffraction limit, with roughly equal probabilities, anywhere within a distance

$$\Delta x_{\text{measure}} \simeq \frac{1}{6}\lambda\frac{L_1}{a} \tag{1.7}$$

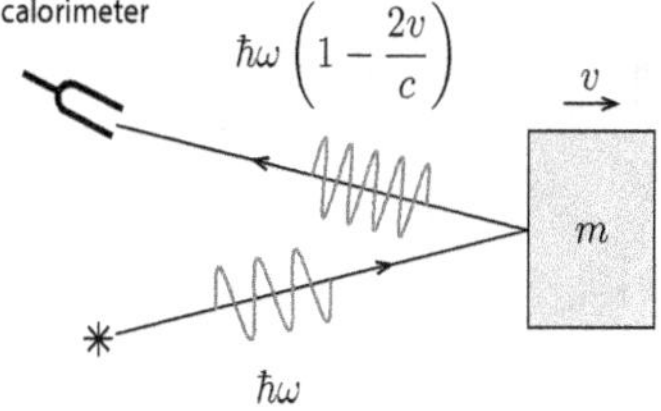

Figure 1.2: Schematic of von Neumann's Doppler speed meter experiment.

of the location x_1. Because the photon has momentum $p = h/\lambda$, it therefore must have given the stick a *random* momentum in the x direction

$$\Delta p_{\text{perturb}} \geq \frac{\hbar\omega}{c}\frac{a}{2L_1}. \tag{1.8}$$

The product of Eqs. (1.7) and (1.8) gives the Heisenberg uncertainty relation

$$\Delta x_{\text{measure}}\Delta p_{\text{perturb}} \geq \frac{\hbar}{2}. \tag{1.9}$$

von Neumann's Doppler speed meter

Another famous thought experiment to demonstrate the Heisenberg uncertainty relation is the von Neumann's Doppler speed meter [ ]. In contrast to the Heisenberg's microscope experiment which measure the object's position, this experiment considers the measurement of the object's speed or momentum. As shown in Fig. (1.2), we shoot a monochromatic single photon pulse with duration τ (and therefore frequency range $\Delta\omega \simeq 1/\tau$ due to the wave properties) to the macroscopic object with mass m and speed v, we then measure the energy of the reflected photon with a calorimeter. Because of the Doppler effect, the *mean* frequency of the photon will change by the amount

$$\frac{\delta\omega}{\omega} = \frac{2v}{c}. \tag{1.10}$$

However, the initial energy was known only with a fractional error $\Delta\omega/\omega = 1/\omega\tau$, and therefore the precision of the object's speed will be within

$$v_{\text{measure}} = \frac{c}{2}\frac{\Delta\omega}{\omega} = \frac{c}{2\omega\tau}. \tag{1.11}$$

The photon's reflection gives the body a precisely known momentum $2\hbar\omega/c$, but the momentum of time when the momentum is transferred is known up to the pulse duration τ. Therefore, the object's position is perturbed by

$$\Delta x_{\text{perturb}} \geq \frac{2\hbar\omega}{mc}\frac{\tau}{2}. \tag{1.12}$$

The product of the uncertainties (1.8) and (1.7) again gives the Heisenberg uncertainty relation

$$\Delta x_{\text{perturb}}\Delta p_{\text{measure}} \geq \frac{\hbar}{2}, \tag{1.13}$$

where $\Delta p_{\text{perturb}} = m\Delta v_{\text{measure}}$. Compare to the uncertainty relation (1.9), the roles of position and momentum in (1.13) are interchanged.

Discussion

As pointed out by Braginsky et al [ ], although the uncertainty relations (1.6) and (1.9) look similar, their essences are fundamentally different. The uncertainty relation (1.6) is the result of the fundamental properties of the quantum object. It simply tells us what a quantum object has to be, i.e., a quantum object doesn't have precisely defined values of position and momentum *simultaneously*. On the other hand, the uncertainty relation (1.9) is the fundamental property of the measurement process. In this case, the position uncertainty is an error in the measurement, an intrinsic property of the the measurement apparatus, which is in principle independent of the measurand. The momentum uncertainty is the perturbation given to the object by the measuring process, which is determined by the interaction between the quantum object and the measurement apparatus.

While the Heisenberg uncertainty relations (1.6) and (1.9) seem to have different physical origins. In fact, the physical roots of them are the same: note that the measurement apparatus also follows the uncertainty principle (1.6) (in this case, the photon), and the consequence of these unavoidable uncertainties is the random momentum perturbation to the measurand in the manner of (1.9). On the other hand, the unavoidable random perturbation accompanying with the measurement prevents one from prepare a state or obtain initial conditions that violate the uncertainty principle (1.1). Therefore, the two uncertainty relations are complementary to each other.

Quantum limit on repeated measurement

As shown in the Heisenberg's microscope experiment and the von Neumann's Doppler meter experiment, due to the Heisenberg uncertainty principle, any attempt to extract information from an object would inevitably perturb it in an *unpredictable* way. The immediate consequence of this unavoidable random perturbation is that the measurement process can affect the outcomes of the subsequent measurement results. Consider measuring the position of a free mass at time $t = 0$ (e.g., with the Heisenberg microscope) with precision $\Delta x\,(0)$, which would perturb the momentum by the amount $\Delta p\,(0)$. After a period of time τ, the accuracy of a second position measurement is spoiled by the momentum perturbation induced by the first measurement, which is

$$[\Delta x\,(\tau)]^2 = [\Delta x\,(0)]^2 + \left[\Delta p\,(0)\,\frac{\tau}{m}\right]^2 \geq [\Delta x\,(0)]^2 + \left[\frac{1}{\Delta x\,(0)}\frac{\hbar\tau}{2m}\right]^2 \geq \frac{\hbar\tau}{m}. \qquad (1.14)$$

As shown in Eq. (1.14), if we made the initial position measurement very precise ($\Delta x\,(0) \to 0$), then the momentum perturbation would be very big ($\Delta p\,(0) \to \infty$), and as a result it would ruin the precision of the subsequent position measurement ($\Delta x\,(\tau) \to \infty$). The minimum of the position uncertainty $\Delta x\,(\tau)$ occurs at

$$\Delta x\,(0) = \Delta x_{\mathrm{SQL}} \equiv \sqrt{\frac{\hbar\tau}{2m}} \qquad (1.15)$$

and

$$\Delta p\,(0) = \Delta p_{\mathrm{SQL}} \equiv \sqrt{\frac{\hbar m}{2\tau}}, \qquad (1.16)$$

where x_{SQL} and p_{SQL} are the standard quantum limits of a free mass.

Another example is measuring the position of a harmonic oscillator. similar to the previous example, the measure of the position with precision $\Delta x\,(0)$ would produce a minimum momentum perturbation $\Delta p\,(0) = \hbar/2\Delta x\,(0)$. After a period of time τ, the precision of the second position measurement is

$$[\Delta x\,(\tau)]^2 = [\Delta x\,(0)]^2 \cos^2 \omega\tau + \left[\frac{\Delta p\,(0)}{m\omega}\right]^2 \sin^2 \omega\tau$$

$$\geq [\Delta x\,(0)]^2 \cos^2 \omega t + \left[\frac{\hbar}{2m\omega\Delta x\,(0)}\right]^2 \sin^2 \omega t, \tag{1.17}$$

where ω is the frequency of the harmonic oscillator. The minimum rms position error is obtained at

$$\Delta x\,(0) = \Delta x_{\text{SQL}} \equiv \sqrt{\frac{\hbar}{2m\omega}} \tag{1.18}$$

and

$$\Delta p\,(0) = \Delta p_{\text{SQL}} \equiv \sqrt{\frac{\hbar m\omega}{2}}, \tag{1.19}$$

where x_{SQL} and p_{SQL} are the standard quantum limits of a harmonic oscillator.

At the first glance, one might think that the roles of position and momentum are interchangeable in this case, as in the case of the Heisenberg's microscope and the von Neumann's Doppler meter. Surprisingly, this is not the case. Consider measuring the momentum (or velocity) of the free mass (e.g., with the von Neumann's Doppler speed meter) with precision $\Delta p\,(0)$ at $t = 0$: according to the Heisenberg uncertainty relation, it would perturb the position of the mass by at least the amount $\Delta x\,(0) = \hbar/2\Delta p\,(0)$. However, since the momentum is a constant of motion of a free mass, the perturbation of the position wouldn't spoil the momentum during free evolution, and the uncertainty of the momentum remains the same, i.e.,

$$[\Delta p\,(\tau)]^2 = [\Delta p\,(0)]^2. \tag{1.20}$$

Therefore, on can in principle measure the momentum of a free mass with infinitely accuracy $(\Delta p\,(t) \to 0)$ without affecting the results of the subsequent momentum measurement. This type of measurement is called quantum nondemolition (QND) measurement, and we will return to this subject in section (2.3).

1.2 Historical review

In the early days of the twentieth century, experiments were limited to a single measurement of systems that consisted of huge numbers of microscopic objects, probably due to the lack of technologies to perform repeated measurements with sufficient precision and individually manipulate a single microscopic object. These experiments revealed the wave properties of matters, and therefore the Heisenberg uncertainty principle (1.1), in a statistical sense. These type of experiments can be well explained by calculation based on time-dependent perturbation theory [ ] (or Fermi golden rule [ ]), which is based on the "repeated random phase assumption". It assumes that the phase relations between wave functions at the initial state and the final state are randomized, or equivalently that the initial and final density matrix are always diagonal in the unperturbed basis. Therefore, one only needs to care about the probabilities at the end of the interaction, which are given by the squares of the wave functions. Albeit the lack of rigorous justification of this assumption, these calculations explain the experimental observations successfully. During this time, the effects of measurement due to the Heisenberg principle (1.9) are only conisdered in thought experiments.

The arrival of maser in 1953 [ ] and laser in 1960 [ ] completely changed the situation. These highly coherent light sources can generate electromegnetc fields with high spatial and temperal coherence, in which the repeated random phase assumption doesn't hold, whcih inspires people to develope a quantum theory of light, and gives rise to the field of quantum optics. The ability to prepare highly coherent states and amplify signals at a single photon level makes the previous thought experiments realizable, and stimulates researches on the consequences of the Heisenberg uncertainty principle on linear amplification [, ,] and quantum mechanical measurement [ ]. Furthermore, the call for the detection of the gravitational wave in 1960, which required continuously monitoring the motion of a macroscopic object with the sensitivities close to the quantum level, necessitated the development of the quantum theory of continuous measurement [, , ,] because the traditional quantum theory of measurement didn't treat the continuous case.

In addition to the development of laser, another scientific and technological breakthrough in the late 20th century, which is visioned by Richard Feynman in his famous talk entitled "There's Plenty of Toom at the Bottom" in 1959 [ ], is nanotechnology. By shrinking down the size of structures or materials to nanoscale, many interesting and useful properties appear because the feature size is of the same order as the critical size for the physical phenomena.

Nanotechnology is a very broad notion which includes various fields of science, and here we focus on nano-electromechanical systems (NEMS). Mechanical structures on the nanometer scale have several attractive properties: first, the mass of nanomechanical structures can be very small ($\sim 10^{-21}$ g) [ ]. Second, by carefully designing the mechanical strucrure, the resonance frequency of a nanomechancial resonantor can cover various range from few kHz to few GHz [, , , , ,]. Third, the quality factor of a nanomechancial resonator can be very high, as high as 10^8 at room temperature [ ]. All these properties together make the nanomechancial resonator extremely susceptible to external alterations. It has been applied for various types of sensitive detection, for example, charge sensing [ ], infrared thermal sensor [ ], inertial sensing [, ,] and imaging [ ], magnetic resonance imaging [ ], etc.

In additional to technological applications, the unique properties of the nanomechanical system also offer an opportunity to study quantum mechanics in a massive macroscopic system. Since the frerquency of the nanomechancial resonator can be as high as GHz, at cryogenic temperature (~ 10 mK), the mechancial resonator occupies it's quantum ground state. By tighly coupling it with a superconducting qubit as a detector, O'Connell et al. have demonstrated the quantum motion of a macroscopic mecahnical object [ ]. By reversing the role of the nanomechanical resonator and the qubit, LaHaye et al. have demonstrated a nanomechanical read out of a superconducting qubit [ ].

The ability to tighly couple a nanomechancial resonator to various mesoscopic quantum systems also enables us to continuously monitor the mechanical motion close to the standard quantum limit (SQL), which brings the interest of quantum measurement into nanomechanical systems and stimulates a lot of research on the quantum limit of detection of a nanomechancial resonator with various detection techniques [,   ,   ]. A seminal work is given by LaHaye et al.: by coupling a nanomechancial resonator to a superconducting single-electron transistor (SSET), they demonstrated a continuous detection of position with resolution close to the SQL [   ]. Later, by applying a microwave cavity interferrometer based detector, and incorporating a near quantum-limited microwave amplifier, Teuful et al. demonstrated a position measurement with measurement imprecision below the SQL [   ]. In addition to continuous position detection where the measurement precision is limited by the SQL, by carefully modulating the coupling between the nanomechanical resonator and a cavity in the resolved sideband regime (mechanical resonance frequency much larger than the cavity linewidth), one can perform QND measurement of the mechanical quadrature [,   ]. By applying this technique, Suh et al. have demonstrate a backaction-evading measurement of a single mechanical quadrature with measurement imprecision below the SQL [   ].

Besides detection, the backaction from the detector can generate dynamical effects to the mechanical motion [   ], and the dynamical backaction from the detector can be applied to control the motion of the nanomechanical resonator. In particular, in sideband resolved cavity opto/electro-mechanical systems [,   ,   ,   ,   ,   ], where the mechanical motion is tighly coupled to the cavity resonance frequency, the high Q cavity enables one to selectively enhance the mechanical sidebands, and therefore the backaction from the Stokes and anti-Stokes scattering processes. Reviews of the filed of opto/electromechanics can be found in [,   ]. By applying the sideband cooling technique in the opto/electro-mechancial systems [,   ], one can cool the motion of various nanomechanical resonators into the quantum ground state [,   ,   ]. The ability to actively cool the mechanics into the ground state brings nanomechancial systems into quantum regime: the zero-point motions of mesoscopic mechanical resonators have been observed [,   ,   ,   ], the quantum radiation shot noises have been detected with nanomechanical resonators [,   ], quantum squeezed states of light have been generated with optomechanical systems [,   ], entanglement between the microwave fields and the mechanical motion has been generated [   ], and recently quantum squeezed states of mesoscopic mechanical resonators have been generated [,   ,   ,   ] by means of the dynamical backaction [   ]. The combination of nanomechanical resonators with high quality microresonators and other quantum systems such as superconducting qubit opens up a lot of opportunities for practical applications [,] as well as fundamental study of quantum mechanics [].

1.3   Thesis overview

Chapter 2 discusses the consequences of the Heisenberg uncertainty principle to measurement and the quantum state of motion. The quantum limit on linear amplification and continuous position detection are analyzed. Quantum non-demolition (QND) measurement will be introduced to go beyond the quantum limit. At the end, an introduction of the minimum uncertainty states will be discussed.

Chapter 3 describes how to implement a cavity electromechancial system out of electrical circuit and nanomechanical resonator. In this chapter, both classical and quantum descriptions of a cavity electromechanical system are given. The quantum Langevin equations of a optomechanical system will be derived. Connection between the language of microwave engineering to that of quantum optics will be discussed.

Chapter 4 analyzes a cavity electromechanical system under bichromatic drives with the quantum Langevin equations derived in chapter 3. General solutions of the transmission spectrum and noise spectra will be calculated. The backaction effects from the cavity fields and the measurement precision of some special drive configurations are analyzed.

Chapter 5 provides the details of the experimental setup. Detailed of the design and fabrication processes for the cavity electromechanical system will be discussed. The measurement techniques and cryogenic setup in the experiments will be presented.

Chapter 6 implements the backaction-evading (BAE) measurement of a single mechanical quadrature with the cavity electromechanical system. In this experiment, we demonstrate a single quadrature measurement with imprecision below the standard quantum limit, while evading the measurement backaction at the same time. Moreover, by incorporating an additional BAE setup, we observe the quantum radiation backaction force from the quantum fluctuation of microwave field in the cavity.

Chapter 7 implements the two-tone reservoir engineering technique with the cavity electromechanical system to squeeze the mechanical motion. Using an independent BAE measurement to directly quantify the squeezing, we observe more than 3 dB of squeezing below the zero-point level, surpassing the 3 dB limit of standard parametric squeezing techniques. Our measurements also reveal evidence for an additional mechanical parametric effect. The interplay between this effect and the optomechanical interaction enhances the amount of squeezing obtained in the experiment.

Chapter 2

Theoretical background

2.1 Quantum measurement

In the last chapter, we introduced the process of measurement in quantum mechanics with the thought experiments introduced by Heisenberg and von Neumann. Here we will continue the discussion of measurement in quantum mechanics. Here we adopt the definition of measurement by Landau and Lifshitz [ ]: "By measurement, in quantum mechanics, we understand any process of interaction between classical and quantum objects, occurring apart from and independently of any observer." Here we call the "classical object" apparatus, which is a physical object that obeys classical mechanics to a sufficient degree of accuracy, and its state would be altered after it interacts with the quantum object. Examples of the classical measuring devices are the photoemulsion of a photographic plate or the supersaturated steam in a Wilson cloud chamber. The classical apparatus usually consists of many degrees of freedom and contains considerable randomness. It follows the law of thermodynamics and therefore is in principle irreversible; the perturbation induced by the classical apparatus is far stronger than the minimum perturbation required by the Heisenberg uncertainty relation (1.9). In general, there are two ways to perform measurement on a quantum system: direct measurement and indirection measurement [ ].

2.1.1 Direct measurement and indirect measurement

The most straightforward way to measure a quantum system is to let the system directly interact with the classical apparatus, which is called "direct measurement". For example, measuring the traces of a particles with a Wilson chamber, measuring the motion of microparticles in the photoemulsion, ... However, as described earlier, the classical apparatus is a pair of filthy hands: it collapses the measurand wave function and applies strong perturbation to the measurand directly.

A better measurement can be made by "indirect measurement", examples of which are Heisenberg's microscope and von Neumann's speed meter. Instead of touching the intricate measurand with the filthy hands (the classical apparatus, e.g., the photographic plate in the Heisenberg's microscope and the calorimeter in von Neumann's speed meter) directly, a quantum proxy (e.g., photon in the thought experiments) is inserted in between the measurand and the classical apparatus, which is a quantum system that has been prepared in advance in some special initial quantum state. In indirect measurement, the measurand first interacts with the proxy to generate correlation between the measurand state and the quantum proxy. Next, a direct measurement of some chosen observable of the proxy is performed. The direct measurement reduces the state of the proxy and therefore the state of the measurand due to the correlation generated in the first step. As a result, the strong perturbation from the classical apparatus applies to the proxy instead of the measurand, and thus the perturbation applies to the measurand by the quantum proxy remains quantum limited.

Note that in order to reach quantum limited precision in an indirection measurement, the experimental apparatus (the classical apparatus and the quantum proxy) has to satisfy two conditions: First, the second step (interaction between the proxy and the classical apparatus) of the measurement should not begin until the first step (interaction between the measurand and the proxy) is complete. This condition isolates the measurand from the "killing-of-the-quantum-state" influence of the classical device. Second, the second step shouldn't contribute significantly to the total error of the measurement. For example, the finite size of the silver grain should be smaller than the diffraction limit of the lens.

2.1.2 Continuous linear measurement

2.1.2.1 Linear measurement

Among the various classes of measurements, the most important class is the linear one. Linear measurements are particularly important due to their simplicity and tight connection to linear systems, in which the equations of motion for the related physical quantities are linear, such as the position and momentum of a free particle or a harmonic oscillator.

This class of measurement can be fully described by a set of linear equations, for example, in the case of the Heisenberg microscope,

$$\tilde{x} = x_{\text{init}} + \delta x_{\text{measure}}, \tag{2.1}$$

$$p = p_{\text{init}} + \delta p_{\text{perturb}}. \tag{2.2}$$

Here $\tilde{x}$ is the result of the coordinate measurement, and p is the object's momentum just after the measurement. x_{init} and p_{init} are the coordinate and momentum just before the measurement, which depend on the initial state of the object. $\delta x_{\text{measure}}$ and $\delta p_{\text{perturb}}$ are the imprecision and random perturbation from the measurement. Linear measurement require that $\delta x_{\text{measure}}$ and $\delta p_{\text{perturb}}$ are statistically independent of the object's initial state, and satisfy the uncertainty relation (1.9) and depend only on the initial state of the probe. In other words, linear measurement required that the uncertainty relations for the observables and its perturbed variables are state dependent, and a counter example is the spin of a spin-1/2 particle, where the uncertainty relation of its spin in the x and y directions is

$$\Delta S_x \Delta S_y \geq \frac{\hbar}{2} |\langle S_z \rangle|, \tag{2.3}$$

where S_x, S_y, and S_z are the spin in x, y, and z directions. Unlike the case of position and momentum of a particle, in this case, the uncertainty relation of S_x and S_y dependsP on the state of the particle.

2.1.2.2 Continuous linear measurement

Up to this point, we have been considering the effects of the uncertainty principle in a single measurement or repeated measurement. In real world application, continuous linear measurement is of particular interest. The interest of continuous linear measurement arose in the 1960s, when the technologies (masers [ ], lasers [,], parametric amplifiers [ ], ...) of making continuous linear measurements get close to the level where the quantum effects have to be taken into account. However, the traditional quantum theory of measurement did not treat continuous measurements. Different approaches to the analysis of continuous quantum measurement werte developed during the 1980s [, , ,]. Here we follow the approach of Braginsky et al. [ ] to discuss the consequences of the Heisenberg uncertainty principle in continuous linear measurement.

To begin the discussion, let's go back again to the Heisenberg' microscope experiment. Now, instead of a single measurement of the object's position, we track the position of the object with a series of measurements with time interval θ, which is sufficiently short such that the object's position x will not change substantially between measurements. In this case, we can average the results of several successive measurements to improve the measurement precision. If n measurements are averaged, then the position will be known with the precision

$$\Delta_\tau x = \frac{\Delta x_{\text{measure}}}{\sqrt{n}} = \Delta x_{\text{measure}} \sqrt{\frac{\theta}{\tau}}, \tag{2.4}$$

where $\tau = n\theta$ is the averaging time, which is chosen to be much shorter than the interested time scale of the object. $\Delta x_{\text{measure}} \propto \lambda$ is the imprecision of a single measurement, which is given by Eq. (1.7) in the case of Heisenberg's microscope.

Now if we reduce the interval θ between measurements to zero, as a consequence of the uncertainty principle of photon (1.3), the imprecision of each measurement will approach to infinity. To illustrate this point, let's rewrite the uncertainty relation (1.3) to be

$$\theta \Delta \lambda \geq \frac{\lambda_0^2}{2c}, \tag{2.5}$$

where $\theta = \Delta x/c$. In the previous discussion of the Heisenberg's microscope, the wavelength of the photon is assumed to have a precise value, i.e., $\Delta\lambda = 0$, and according to the uncertainty relation (2.5) it requires infinite time to perform the measurement, i.e., $\theta \to \infty$ and the imprecision of the position measurement is given by (1.7). Now, as the measurement time $\theta \to 0$, the uncertainty of the photon's wavelength $\Delta\lambda \to \infty$, and therefore the uncertainty of the object's position $\Delta x_{\text{measure}} \to \infty$. Since the average position imprecision (2.4) and the average time τ are finite, the quantity

$$S_x \equiv (\Delta x_{\text{measure}})^2 \theta \tag{2.6}$$

remains finite. The precision during a fixed averaging time τ can be expressed as

$$\Delta_\tau x = \sqrt{\frac{S_x}{\tau}}. \tag{2.7}$$

To clarify the physical meaning of S_x, let's consider the continuous version of (2.1), which is

$$\tilde{x}(t) = x(t) + x_{\text{fluct}}(t), \tag{2.8}$$

where $\tilde{x}(t)$ is the output signal of the measuring device, which is given by the sum of the input signal $x(t)$ and the noise $x_{\text{fluct}}(t)$ added by the apparatus. The quantity S_x is the spectral density of the added noise $x_{\text{fluct}}(t)$.

Next, let's consider the backaction of the measuring device on the object. The continuous version of (2.2) is

$$p_{\text{after}}(t) = p_{\text{before}}(t) + p_{\text{perturb}}(t), \tag{2.9}$$

where $p_{\text{before}}(t)$ and $p_{\text{after}}(t)$ is the object's momentum right before and right after the measurement. $p_{\text{perturb}}(t)$ is the random momentum perturbation induced by the measurement, the rms change of this momentum perturbation $\Delta P_{\text{perturb}} \propto 1/\lambda$, which is given by Eq. (1.8) for Heisenberg's microscope. The variance of the momentum increases in the manner of a diffusive process, and after the averaging time τ, the rms momentum perturbation is

$$\Delta_\tau p = \Delta p_{\text{perturb}} \cdot \sqrt{n} = \Delta p_{\text{perturb}} \cdot \sqrt{\frac{\tau}{\theta}}. \tag{2.10}$$

as $\theta \to 0$ and $\lambda \to \infty$, the collisions of the object with the photons become more frequent and the change of momentum per collision becomes smaller in such a way that the rms deviation of the momentum after a fixed time remains constant. We can define a finite quantity

$$S_F = \frac{1}{\theta} \left(\Delta p_{\text{perturb}} \right)^2 \tag{2.11}$$

and Eq. (2.10) can be expressed as

$$\Delta_\tau p = \sqrt{S_F \tau}. \tag{2.12}$$

The quantity S_F is the spectral density of the backaction force.

By taking the product of Eqs. (2.6) and (2.6) and using the Heisenberg uncertainty relation (1.9), we obtain

$$S_x \cdot S_F = (\Delta x_{\text{measure}})^2 \cdot \left(\Delta p_{\text{perturb}} \right)^2 \geq \frac{\hbar^2}{4}. \tag{2.13}$$

This inequality plays the same role as the Heisenberg relation (1.9) in discrete measurement, and it establishes a universal, mutual connection between the measurement imprecision and the measurement backaction. A general and rigorous analysis can be found in [ ].

2.2   Quantum limit of amplification and detection

In the previous section, we have discussed the effects of the Heisenberg uncertainty principle to measurement by considering examples such as Heisenberg's microscope and von Neumann's speed meter. In this section, we will discuss the quantum limit of measurement in a general and rigorous context. We will first discuss the approach introduced by Haus et al. [ ] and further clarified and extended by Caves [ ], which focuses on the quantum limit in linear amplifiers of bosonic modes. Then, a more general theory for generic linear amplifier by Clerk [,] will be discussed.

38

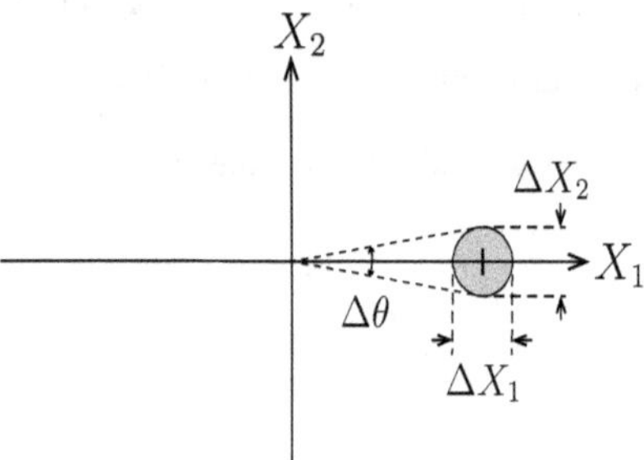

Figure 2.1: Signal in the quadrature space.

2.2.1 Standard Haus-Caves quantum limit

Basic

Consider a narrow band linear amplifier that both of its input and output signals are narrow band signal with bandwidth Δf, i.e.,

$$\hat{x}(t) = \hat{a}e^{-i\omega t} + \hat{a}^{\dagger}e^{i\omega t} = \hat{X}_1 \cos\omega t + \hat{X}_2 \sin\omega t, \tag{2.14}$$

where $\hat{a}$ and $\hat{a}^{\dagger}$ are the annihilation and creation operators of the bosonic mode with carrier frequency ω. The signal information is carried by the two slow varying quadratures $\hat{X}_1(t)$ and $\hat{X}_2(t)$ on a timescale $\tau = 1/\Delta f \gg 2\pi/\omega$. As shown in Fig. 2.1, the information of the signal can be represented in the quadrature space. To illustrate the physical meaning of the quadratures, we choose $\langle \hat{X}_2 \rangle = 0$ without loss of generality. As shown in figure, the uncertainty in X_1 corresponds to amplitude fluctuation of fractional size $\Delta X_1/\langle \hat{X}_1 \rangle$, and the uncertainty in X_2 corresponds to phase fluctuations of size $\Delta\theta = \Delta X_2/\langle \hat{X}_1 \rangle$. According to quantum mechanics, the only non-zero commutator of the bosonic operator $\hat{a}$ and $\hat{a}^{\dagger}$ is (see section 3.4 for detail derivation)

$$\left[\hat{a}, \hat{a}^{\dagger}\right] = 1, \tag{2.15}$$

and therefore

$$\left[\hat{X}_1, \hat{X}_2\right] = 2i, \tag{2.16}$$

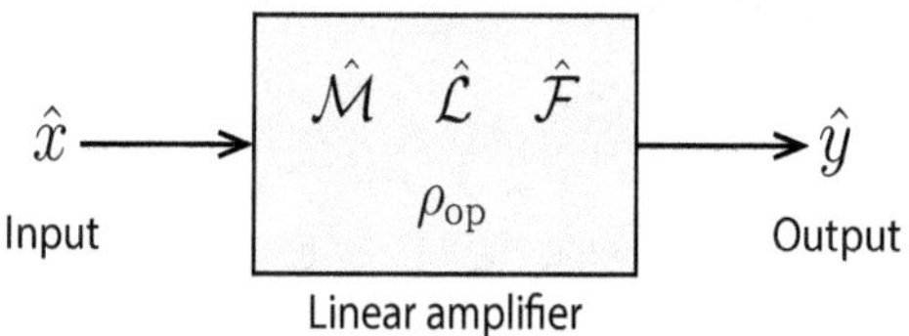

Figure 2.2: Schematic of a linear amplifier.

or in terms of the uncertainty relation, $\Delta X_1 \cdot \Delta X_2 \geq 1$, quantum mechanics place a fundamental limit to the area of the noise blub in Fig. 2.1. In the following, we will discuss the consequences of the uncertainty principle to the added noise of a linear amplifier.

In the following discussion, the output signal of the linear amplifier is represented by

$$\hat{y}(t) = \hat{b}e^{-i\omega t} + \hat{b}^{\dagger}e^{i\omega t} = \hat{Y}_1 \cos \omega t + \hat{Y}_2 \sin \omega t, \tag{2.17}$$

and the input signal is given by Eq. (2.14), where a schematic of the a linear amplifier is given by Fig. 2.2. The output signal is linearly related to the input signal in a linear amplifier, i.e.,

$$\hat{b} = \hat{\mathcal{M}}\hat{a} + \hat{\mathcal{L}}\hat{a}^{\dagger} + \hat{\mathcal{F}}, \tag{2.18}$$

where $\hat{\mathcal{M}}$, $\hat{\mathcal{L}}$, and $\hat{\mathcal{F}}$ are operators that depend only on the internal modes of the amplifier, and therefore they commute with the input mode operators $\hat{a}$ and $\hat{a}^{\dagger}$.

Assumptions

First, we prepare the amplifier into an initial state which is independent of the input state, and the initial density operator of the entire system can be written into

$$\rho = \rho_I \cdot \rho_{op}, \tag{2.19}$$

where ρ_I is the initial density operator for the input state and ρ_{op} is the initial density operator of the amplifier's internal modes. Here we work in the Heisenberg picture, where the density operator does not change with time.

Then, we expand the operators of the internal modes around its operating state, i.e.,

$$\hat{M} = \langle \hat{M} \rangle_{\mathrm{op}} + \delta\hat{M}, \tag{2.20}$$

$$\hat{L} = \langle \hat{L} \rangle_{\mathrm{op}} + \delta\hat{L}, \tag{2.21}$$

$$\hat{\mathcal{F}} = \langle \hat{\mathcal{F}} \rangle_{\mathrm{op}} + \delta\hat{\mathcal{F}}, \tag{2.22}$$

$$\tag{2.23}$$

where $\delta\hat{M}$ and $\delta\hat{L}$ are the gain fluctuations, which introduce multiplicative noises to the output signal. $\delta\hat{\mathcal{F}}$ is the additive noise to the output signal. In the following discussion, we assume the multiplicative noises $\delta\hat{M}$ and $\delta\hat{L}$ are negligible in the operating state, and we set the constant shift of the output signal $\langle \hat{\mathcal{F}} \rangle_{\mathrm{op}} = 0$ without loss of generality. Then, the linear relation (2.18) becomes

$$\hat{b} = M\hat{a} + L\hat{a}^{\dagger} + \hat{F}, \tag{2.24}$$

where $M = \langle \hat{M} \rangle_{\mathrm{op}}$, $L = \langle \hat{L} \rangle_{\mathrm{op}}$ are the expectation values of the gains, and $\hat{F} = \delta\hat{\mathcal{F}}$ is the additive noise of the amplifier.

Uncertainty relations and quantum limit on added noise

Since both the input mode and the output mode follow the bosonic commutation relation (2.15), which implies

$$\left[\hat{F}, \hat{F}^{\dagger} \right] = 1 - |M|^2 + |L|^2, \tag{2.25}$$

or in terms of the uncertainty relation

$$|\Delta F|^2_{\mathrm{op}} \geq \frac{1}{2} \left| 1 - |M|^2 + |L|^2 \right|. \tag{2.26}$$

We can rearrange the linear relation (2.24) in terms of quadratures

$$\hat{Y}_1 = (M + L)\, \hat{X}_1 + \hat{F}_1, \tag{2.27}$$

$$\hat{Y}_2 = (M - L)\, \hat{X}_2 + \hat{F}_2, \tag{2.28}$$

where

$$\hat{F}_1 = \frac{1}{2}\left(\hat{F} + \hat{F}^\dagger\right), \tag{2.29}$$

$$\hat{F}_2 = -\frac{i}{2}\left(\hat{F} - \hat{F}^\dagger\right). \tag{2.30}$$

The uncertainty relation (2.25) implies

$$\left[\hat{F}_1, \hat{F}_2\right] \geq \frac{i}{2}\left(1 - |M|^2 + |L|^2\right), \tag{2.31}$$

or in terms of uncertainty relation

$$\Delta F_1 \cdot \Delta F_2 \geq \frac{1}{4}\left|1 - |M|^2 + |L|^2\right|. \tag{2.32}$$

The fluctuations of the output quadratures are

$$(\Delta Y_i)^2 = G_i\,(\Delta X_1)^2 + (\Delta F_i)^2, \tag{2.33}$$

where $i = 1, 2$, the quadratures' power gain $G_1 = (M + L)^2$, and $G_2 = (M - L)^2$. The added noise is conveniently characterized by the *added noise numbers*

$$A_i \equiv |\Delta F_i|^2 / G_i, \tag{2.34}$$

and the uncertainty relation of the added noises directly follow the commutation relation (2.31), which is

$$\sqrt{A_1 A_2} \geq \frac{1}{4}\left|1 \mp \frac{1}{\sqrt{G_1 G_2}}\right|, \tag{2.35}$$

where the upper (lower) sign holds if $|M| \geq |L|$ ($|M| \leq |L|$). The uncertainty relation implies that, as a general rule, a reduction in the noise added to one quadrature phase requires an increase in the noise added to the other phase.

Phase-insensitive linear amplifier

To illustrate the quantum limit of the added noise, let's first consider a phase-insensitive linear amplifier, which amplifies both quadratures equally, i.e., $G_1 = G_2 = G = M^2$ (phase preserving) or $G_1 = G_2 = G = L^2$ (phase conjugating), and $A_1 = A_2 = A/2$. Then the uncertainty relation (2.35) becomes

$$A \geq \frac{1}{2}\left|1 \mp \frac{1}{G}\right|. \tag{2.36}$$

Therefore, a high gain phase-insensitive amplifier must added at least half quanta of noise to both quadratures at the input signal. Note that if there is no gain ($G = 1$), the amplifier need not add any noise.

Phase-sensitive linear amplifier

For phase-sensitive linear amplifier, one can surpass this quantum limit (2.36) in one of the quadrature. For example, consider an amplifier such that $M^2 - L^2 = \sqrt{G_1 G_2} = G \gg 1$, and therefore $A_1 \cdot A_2 \geq \frac{1}{4}\left|1 - \frac{1}{G}\right| \to \frac{1}{4}$. One can design the amplifier so that it amplifies the quadrature of interest, for example $G_1 \gg 1$, and the added noise for that quadrature can be below the quantum limit for the phase-insensitive amplifier ($A_1 \ll 1/4$) in the expense of increasing the added noise to the other quadrature ($A_2 \gg 1/4$). This type of phase-sensitive amplifier can be realized with degenerate parametric process. Another example is given by designing an amplifier such that $M = L = \sqrt{G}/2 \gg 1$ ($G_1 = G$ and $G_2 = 0$). The amplifier completely give up the information in the X_2 quadrature, in this case $\Delta F_1 \cdot \Delta F_2 \geq 1/4$. Therefore, the added noise in the X_1 quadrature can be below the quantum limit when the gain is sufficiently high ($A_1 = \Delta F_1/\sqrt{G} \ll 1/4$). This is the backaction-evading (BAE) technique we used to monitor the mechanical motion.

2.2.2 Linear response theory

In this section, we will introduce the linear response approach developed by Clerk et al. [,] to discuss the quantum limit of a *generic* linear response position detector.

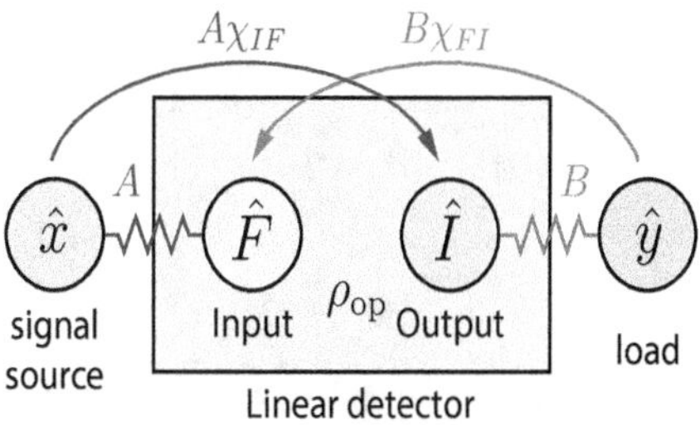

Figure 2.3: Schematic of a linear response detector.

Unlike the Haus-Caves approach where both the input and the output are bosonic modes, this approach considers a generic linear detector which can be a fermonic base such as single electron transistor (SET) or quantum point contact (QPC). Moreover, this approach directly involves the noise properties of the detector, and allows one to derive simple neccessary and sufficient conditions for reaching quantum limit. Note that the linear response approach can be applied to different systems [, ,], and here we focus on the discussion of the quantum limit of continuous position detection.

Basic

In the linear response approach, the detector is a physical system with Hamiltonian $\hat{H}_{\mathrm{det}}$, and it has an input, characterized by the operator $\hat{F}$, and an output, chanracterized by the operator $\hat{I}$, the schematic of the linear response detector is shown in Fig. 2.3. The input port is linearly coupled to the signal operator $\hat{x}$ with the interaction Hamiltonian

$$\hat{H}_{\mathrm{int}} = -A\hat{x} \cdot \hat{F}, \tag{2.37}$$

where $A\hat{F}$ is the backaction applyis the quantity that to the signal source. $\hat{I}$ is the physical quantity that is read out at the output of the detector. In the following, we work in the Heisenberg picture with respect to the detector Hamiltonian $\hat{H}_{\mathrm{det}}$.

We consider that the coupling A is sufficiently weak such that the output of the detectort can be described by the linear response relation

$$\langle \hat{I}(t) \rangle = \langle \hat{I} \rangle_0 + A \int dt' \chi_{IF}(t-t') \langle \hat{x}(t') \rangle, \tag{2.38}$$

where $\langle ... \rangle_0$ indicate the expectation value with respect to the initial density matrix of the uncoupled detector. $\langle I \rangle_0$ is the input-independent value of the detector output at zero coupling. $\chi_{IF}(t)$ is the linear-response susceptibility or gain of the detector, which is given by the Kubo-like formula

$$\chi_{IF}(t) = -\frac{i}{2}\theta(t) \langle [\hat{I}(t), \hat{F}(0)] \rangle_0. \tag{2.39}$$

In general, the detector can operate in reverse, i.e., $\hat{H}_{\text{Int}} = -A\hat{x}\hat{I}$, and

$$\langle \hat{F}(t) \rangle = \langle \hat{F} \rangle_0 + A \int dt' \chi_{FI}(t-t') \langle \hat{x}(t') \rangle \tag{2.40}$$

with reverse gain

$$\chi_{FI}(t) = -\frac{i}{\hbar}\theta(t) \langle [\hat{F}(t), \hat{I}(0)] \rangle_0. \tag{2.41}$$

Uncertainty relation and quantum-ideal detector

Both the input and the output operators are subjected to the unavoidable noise. Here, we quantitatively characterize this noise with the symmetric-in-frequency part of tyhe noise spectral density. Redefining the operators $\hat{F} \rightarrow \hat{F} - \langle F \rangle_0$ and $\hat{I} \rightarrow \hat{I} - \langle I \rangle_0$, the symmetric spectral density of the input and output are

$$\bar{S}_{FF}[\omega] \equiv \frac{1}{2} \int_{-\infty}^{\infty} dt\, e^{i\omega t} \langle \{\hat{F}(t), \hat{F}(0)\} \rangle_0, \tag{2.42}$$

$$\bar{S}_{II}[\omega] \equiv \frac{1}{2} \int_{-\infty}^{\infty} dt\, e^{i\omega t} \langle \{\hat{I}(t), \hat{I}(0)\} \rangle_0, \tag{2.43}$$

$$\bar{S}_{IF}[\omega] \equiv \frac{1}{2} \int_{-\infty}^{\infty} dt\, e^{i\omega t} \langle \{\hat{I}(t), \hat{F}(0)\} \rangle_0, \tag{2.44}$$

where $\bar{S}_{II}$ is the intrinsic noise in the output of the detector, $\bar{S}_{FF}$ is the backaction noise seen by the source of the input signal, and $\bar{S}_{IF}$ is the correlation between the input and output noise.

Following the Heisenberg's uncertainty relation

$$(\Delta A)^2 (\Delta B)^2 \geq \frac{1}{4} \left\langle \{\hat{A}, \hat{B}\} \right\rangle^2 + \frac{1}{4} \left\langle \left[\hat{A}, \hat{B}\right] \right\rangle^2, \tag{2.45}$$

the noise spectral densities have to follow the inequality [,]

$$\bar{S}_{II}[\omega] \bar{S}_{FF}[\omega] - \left|\bar{S}_{IF}[\omega]\right|^2 \geq \left|\frac{\hbar \tilde{\chi}_{IF}[\omega]}{2}\right|^2 \left(1 + \Delta\left[\frac{\bar{S}_{IF}[\omega]}{\hbar \tilde{\chi}_{IF}/2}\right]\right), \tag{2.46}$$

where $\Delta[z] = \left[\left|1 + z^2\right| - \left(1 + |z|^2\right)\right]/2$ and $\tilde{\chi}_{IF}[\omega] \equiv \chi_{IF}[\omega] - \chi_{FI}^*[\omega]$. A *quantum-ideal detector* (at a given frequency ω) is defined as one which minimize the left-hand side of Eq. (2.46). In this work, we focus on detectors without reverse gain, i.e., $\chi_{FI} = 0$. The condition to have a quantum-limited detector becomes

$$\bar{S}_{II}[\omega] \bar{S}_{FF}[\omega] - \left|\bar{S}_{IF}[\omega]\right|^2 = \left|\frac{\hbar \chi_{IF}[\omega]}{2}\right|^2 \left(1 + \Delta\left[\frac{\bar{S}_{IF}[\omega]}{\hbar \chi_{IF}/2}\right]\right). \tag{2.47}$$

This ideal noise requirement places a strong constraint on the properties of the detector. As discussed in [,], for a quantum-limited detector, there must exist a complex number α such that

$$\langle f|I|i\rangle = \alpha \langle f|F|i\rangle, \tag{2.48}$$

where $|i\rangle$ and $|f\rangle$ are the initial and final states that contribute to the noise spectra $\bar{S}_{FF}[\omega]$ and $\bar{S}_{II}[\omega]$. The amplitude and imaginary part of the complex number α are given by

$$|\alpha[\omega]| = \sqrt{\bar{S}_{II}[\omega]/\bar{S}_{FF}[\omega]}, \tag{2.49}$$

$$\frac{\mathrm{Im}\{\alpha[\omega]\}}{|\alpha[\omega]|} = -\frac{\hbar \chi_{IF}[\omega]/2}{\sqrt{\bar{S}_{II}[\omega]\bar{S}_{FF}[\omega]}}. \tag{2.50}$$

Note that in order to have a non-vanishin gain ($\chi_{IF}[\omega] \neq 0$), one needs $\mathrm{Im}\{\alpha[\omega]\} \neq 0$. It implies that the set of all initial states $|i\rangle$ has no overlap with the set of all final states $|f\rangle$, which means that a quantum limited detector cannot be in equilibrium.

Detector backaction

As mentioned in the previous section, the detector would generate backaction force $A\hat{F}$ to the source of the input signal, in this case the mechanical oscillator. In this section, we will discuss the consequence of the backaction. The motion of the mechanical oscillator under detection can be described by an effective classical Langevin equation,

$$m\ddot{x}(t) = -m\omega_m^2 x(t) - m\Gamma_m \dot{x}(t) + F_0(t) - mA^2 \int dt' \Gamma_{\det}(t-t')\,\dot{x}(t') - AF(t), \quad (2.51)$$

where ω_m is the resonance frequency of the mechanical oscillator with mass m. F_0 is the random fluctuating force from the equilibrium phonon bath, whose spectrum is given by the standard equilibrium relation

$$\bar{S}_{F_0 F_0}[\omega] = m\Gamma_m \hbar\omega \coth\left(\frac{\hbar\omega}{2k_B T}\right) = m\Gamma_m \hbar\omega \left(2n_m^T[\omega] + 1\right), \quad (2.52)$$

where $n_m^T[\omega] = 1/\left(e^{\hbar\omega/k_B T} - 1\right)$ is the average mechanical occupation. Γ_m is the damping due to the equilibrium bath, and is related to the asymmetric part of the noise spectrum of F_0 by

$$\Gamma_m = \frac{1}{\hbar m} \frac{S_{F_0 F_0}[\omega] - S_{F_0 F_0}[-\omega]}{2\omega}, \quad (2.53)$$

which is a constant for equilibrium bath. $\Gamma_{\det}$ is the damping induced by the detector. Similar to Γ_m, it is given by the asymmetric part of the backaction force F from the detector, that is

$$\Gamma_{\det}[\omega] = \frac{1}{\hbar m} \frac{S_{FF}[\omega] - S_{FF}[-\omega]}{2\omega}. \quad (2.54)$$

Note that $\Gamma_{\det}$ can be positive or negative depending on the operating state of the detector. Similar to the relation (2.52) in the equilibrium bath, one can define the effective temperature of the detector is

$$\coth\left(\frac{\hbar\omega}{2k_B T_{\mathrm{eff}}[\omega]}\right) \equiv \frac{\bar{S}_{FF}[\omega]}{\Gamma_{\det}[\omega]\,\hbar\omega} = \frac{S_{FF}[\omega] + S_{FF}[-\omega]}{S_{FF}[\omega] - S_{FF}[-\omega]}, \quad (2.55)$$

or interms of symmetric spectrum

$$\bar{S}_{FF}[\omega] = m\Gamma_{\text{det}}[\omega]\,\hbar\omega\,\coth\left(\frac{\hbar\omega}{2k_BT_{\text{eff}}[\omega]}\right) = m\Gamma_{\text{det}}[\omega]\,\hbar\omega\,(2n_{\text{BA}}[\omega]+1),\qquad (2.56)$$

where $n_{\text{BA}}[\omega] = 1/\left(e^{\hbar\omega/k_BT_{\text{eff}}[\omega]} - 1\right)$. Note that the effective temperature $T_{\text{eff}}[\omega]$ simply serves as a measure of the asymmetry of the detector's backaction noise $S_{FF}[\omega]$. As discussed before, a quantum-limited detector cannot be in equilibrium, and one cannot define a physical temperature for a quantum-limited detector. However, as discussed in [,], the power gain of a high gain quantum-limited detector is given by

$$G_p \simeq \left[\frac{\text{Im}\{\alpha[\omega]\}}{|\alpha[\omega]|}\frac{4k_BT_{\text{eff}}[\omega]}{\hbar\omega}\right]^2. \qquad (2.57)$$

In order to have large power gain, one needs to have a large effective detector temperature. Another important consequence of large power gain is that the gain $\chi_{IF}[\omega]$ and the noise cross correlator $\bar{S}_{IF}[\omega]$ are in phase, i.e., $\bar{S}_{IF}[\omega]/\chi_{IF}[\omega]$ is purely real, up to correction that are as small as $\omega/T_{\text{eff}}[\omega]$ [].

The solution of the position fluctuation in frequency space is given by

$$\delta x[\omega] = \chi_{xx}[\omega]\,(F_0[\omega] + AF[\omega]) \qquad (2.58)$$

with the mechanical susceptibility

$$\chi_{xx}[\omega] = -\frac{1/m}{\left(\omega^2 - \omega_m^2\right) + i\omega\Gamma_{\text{tot}}[\omega]}, \qquad (2.59)$$

where $\Gamma[\omega] = \Gamma_m + A^2\Gamma_{\text{det}}[\omega]$ is the total damping rate of the mechanical oscillator. Defining the symmetrized equilibrium position noise of the mechanical oscillator with damping $\Gamma[\omega]$ and average mechanical occupation n_m^{th} by

$$\bar{S}_{xx,\text{eq}}[\omega,T] = \hbar\text{Im}\{\chi_{xx}[\omega]\}\left(2n_m^{\text{th}}+1\right). \qquad (2.60)$$

The symmetric noise spectrum of the mechanical motion is

$$\bar{S}_{xx}[\omega] = |\chi_{xx}[\omega]|^2 \left(\bar{S}_{F_0 F_0}[\omega] + A^2 \bar{S}_{FF}[\omega]\right) = \bar{S}_{xx,\mathrm{eq}}[\omega, \bar{n}_m] \tag{2.61}$$

with the effective phonon occupation

$$\bar{n}_m[\omega] = \frac{\Gamma_m}{\Gamma_{\mathrm{tot}}[\omega]} n_m^T[\omega] + \frac{A^2 \Gamma_{\mathrm{det}}[\omega]}{\Gamma_{\mathrm{tot}}[\omega]} n_{\mathrm{BA}}[\omega]. \tag{2.62}$$

Consider $\omega_m \gg \Gamma_{\mathrm{tot}}$, the only important contributions of $\bar{S}_{F_0 F_0}$ and $\bar{S}_{FF}[\omega]$ is at $\omega = \pm\omega_m$, then Eq. (2.61) becomes symmetric noise spectrum of a mechanical oscillator in an effective Ohmic bath with damping rate $\Gamma_{\mathrm{tot}}[\omega_m]$ and equilibrium phonon occupation $\bar{n}_m[\omega_m]$.

Quantum limit on added noise

After discussed the requirement of a quantum-limited detector and the detector backaction to the mechanical oscillator. In this section, we will derive the quantum limit on the added noise in the position measurement. Here we treat both the position of the mechanical oscillator and the detector output as classically fluctuating quantities. Using the linear relation(2.38), the fluctuation of the detector output in the frequency space is

$$\delta I_{\mathrm{tot}}[\omega] = \delta I_0[\omega] + A\chi_{IF}[\omega]\,\delta x[\omega], \tag{2.63}$$

where δI_0 is the intrinsic fluctuation of the detector output, which has a spectral density $\bar{S}_{II}[\omega]$. Together with the solution (2.58), the spectral density of the total noise in the detector output is given by

$$\bar{S}_{II,\mathrm{tot}}[\omega] = \bar{S}_{II}[\omega] + A^2 |\chi_{IF}[\omega]|^2 \bar{S}_{xx,\mathrm{eff}}[\omega] \tag{2.64}$$

with the effective mechanical spectrum

$$\bar{S}_{xx,\mathrm{eff}}[\omega] = \bar{S}_{xx}[\omega] - 2\mathrm{Re}\left\{\chi_{xx}^*[\omega]\,\bar{S}_{zF}[\omega]\right\}, \tag{2.65}$$

where

$$\bar{S}_{zF}[\omega] = \frac{\bar{S}_{IF}[\omega]}{\chi_{IF}[\omega]}.$$ (2.66)

The spectral density of the position fluctuation inferred from the output of the detector is given by normalizing the detector output by the gain of the detector, i.e.,

$$\bar{S}_{xx,\text{tot}}[\omega] \equiv \frac{\bar{S}_{II,\text{tot}}[\omega]}{A^2 |\chi_{IF}[\omega]|^2} = \frac{\bar{S}_{II}[\omega]}{A^2 |\chi_{IF}[\omega]|^2} + \bar{S}_{xx,\text{eff}}[\omega].$$ (2.67)

To quantify the added noise from the detector, we can rewrite it as

$$\bar{S}_{xx,\text{tot}}[\omega] = \frac{\Gamma_m}{\Gamma_{\text{tot}}[\omega]} \bar{S}_{xx,\text{eq}}\left[\omega, n_m^T\right] + \bar{S}_{xx,\text{add}}[\omega]$$ (2.68)

with the added noise defined by

$$\bar{S}_{xx,\text{add}}[\omega] = \frac{\bar{S}_{II}[\omega]}{A^2 |\chi_{IF}[\omega]|^2} + A^2 |\chi_{xx}[\omega]|^2 \bar{S}_{FF}[\omega] - 2\text{Re}\left\{\chi_{xx}^*[\omega] \bar{S}_{zF}[\omega]\right\}.$$ (2.69)

The first term describes the imprecision of the detector, the second term describes the detector backaction apply to the mechanical oscillator, and the third term describes the correlation between the imprecision and the backaction.

Here we focus on the added noise at the resonance of the mechanical oscillator ($\omega = \omega_m$), in this case, the mechanical susceptibility is purely imaginary, i.e., $\chi_{xx}(\omega_m) = i/m\omega_m\Gamma_{\text{tot}}$. For a quantum-limited detector with high power gain, $\bar{S}_{zF}[\omega]$ is purely real. In this case, the third term in (2.69) becomes zero. Moreover, a quantum-limited detector has to satisfy the noise constraint (2.47), and since there is no further constraint on S_{IF}, we can minimize the expression by setting it equal to zero. Then the added noise (2.69) becomes

$$\bar{S}_{xx,\text{add}} = \frac{\hbar^2}{4} \frac{1}{A^2 \bar{S}_{FF}} + |\chi_{xx}|^2 A^2 \bar{S}_{FF}.$$ (2.70)

Here we consider $\omega = \omega_m$ and neglect the frequency dependence. As shown in (2.70), as one increases the coupling A, the detector imprecision decreases but the detector backaction increases, as shown in Fig. 2.4. In order to obtain the minimum added noise, one needs to

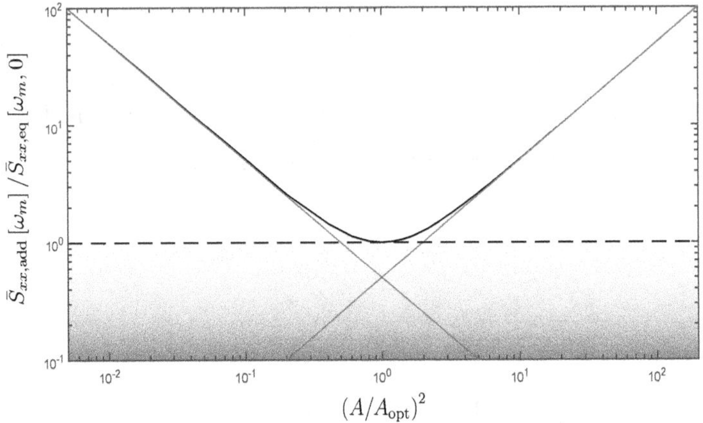

Figure 2.4: Added noise as a function of coupling strength. The blue line is the measurement imprecision, the red line is the measurement backaction, the black curve is the total added noise, and the dashed line represents the standard quantum limit.

attain the optimum coupling

$$A_{\mathrm{opt}}^2 = \frac{\hbar}{2} \frac{1}{|\chi_{xx}| S_{FF}} = \frac{1}{4} \frac{\Gamma_{\mathrm{tot}}}{\Gamma_{\mathrm{det}}} \frac{\hbar\omega_m}{k_B T_{\mathrm{eff}}}. \tag{2.71}$$

At the optimum coupling, the detector imprecision and the detector backaction contribute equally in the added noise, and the minimum added noise of an quantum-limited detector is

$$\left(\bar{S}_{xx,\mathrm{add}}\left[\omega_m\right]\right)_{\min} = \hbar\,|\chi_{xx}\left[\omega_m\right]| = \bar{S}_{xx,\mathrm{eq}}\left[\omega_m, 0\right]. \tag{2.72}$$

We end up with the same result in the Haus-Caves approach for phase-insensitive amplifier, i.e., the added noise of a high power gain quantum-limited position detector must be at least as large as the zero-point noise of the mechanical oscillator. A general derivation for arbitrary ω can be found in [,]. Note that at the optimum couping $A = A_{\mathrm{opt}}$, and the power gain (2.57) at large power gain limit can be expressed as

$$G_p \simeq \left[\frac{\mathrm{Im}\left\{\alpha\right\}}{|\alpha|} \frac{\Gamma_m + A_{\mathrm{opt}}^2 \Gamma_{\mathrm{det}}}{A_{\mathrm{opt}}^2 \Gamma_{\mathrm{det}}}\right]^2, \tag{2.73}$$

which means that in order to achieved quantum limit with large power gain, the intrinsic damping rate of the mechanical oscillator has to be much larger than the damping induced by the detector ($\Gamma_m \gg A_{\mathrm{opt}}^2 \Gamma_{\mathrm{det}}$). Note that if we do not require high power gain, then $\bar{S}_{IF}[\omega]/\chi_{IF}[\omega]$ can be purely imaginary, and the detector does not need to add any noise [,].

2.3 Quantum non-demolition (QND) measurement

In section 1.1, we have shown that not every repeated measurement would generate unpredictable result, for example, momentum measurement with von Neumann's speed meter. This type of measurement that the measurement result is completely predictable from the result of the preceding measurement is called quantum nondemolition (QND) measurement or backaction-evading (BAE) measurement. In this section, we will give a formal discussion on the general conditions of this type of measurement [, ,].

To start the discussion, let's consider an arbitrary quantum mechanical system described by the free system Hamiltonian $\hat{H}_{\mathrm{sys}}$. We measure the physical observable $\hat{A}$ of the system by coupling the observable to a physical apparatus, which is described by the detector Hamiltonian $\hat{H}_{\mathrm{det}}$, with the interaction Hamiltonian $\hat{H}_{\mathrm{Int}}$. The total Hamiltonian for the system plus apparatus is

$$\hat{H} = \hat{H}_{\mathrm{sys}} + \hat{H}_{\mathrm{det}} + \hat{H}_{\mathrm{Int}}. \tag{2.74}$$

A QND measurement of an observable $\hat{A}$ is defined as a sequence of measurements of $\hat{A}$ perform in such a way that the outcomes of each measurement are predictable from the result of the previous measurement.

2.3.1 Requirements on the observable

As described by the definition, QND measurement places a very strong restriction on the observable $\hat{O}$ of the system. Consider performing the first measurement at t_0, according to the projection postulate, the state of the system would be reduced to the normalized eigenstates of $\hat{O}(t_0)$, i.e.,

$$|\psi(t_0)\rangle = \sum_\alpha c_\alpha |O_0, \alpha\rangle. \qquad (2.75)$$

Here we are working in the Heisenberg picture, where α labels the states in any degenerate subspace of $\hat{O}(t_0)$. In order to be a QND measurement, a second measurement will give a predictable result, and the states $|O_0, \alpha\rangle$ must also be eigenstates of $\hat{O}(t_1)$, but not necessarily with the same eigenvalue, i.e.,

$$\hat{O}(t_1)|O_0, \alpha\rangle = f_1(O_0)|O_0, \alpha\rangle, \qquad (2.76)$$

where f_1 is an arbitrary real-value function. This implies the operator equation for a measurement at time t_k,

$$\hat{O}(t_k) = f_k\left[\hat{O}_0\right], \qquad (2.77)$$

or equivalently

$$\left[\hat{O}(t_i), \hat{O}(t_k)\right] = 0. \qquad (2.78)$$

If (2.78) holds only at discrete instants of time, then the observable is called a stroboscopic QND observable. For example, for the position and momentum of a harmonic oscillator,

$$[\hat{x}(t), \hat{x}(t+\tau)] = \frac{i\hbar}{m\omega}\sin\omega\tau, \qquad (2.79)$$

$$[\hat{p}(t), \hat{p}(t+\tau)] = i\hbar m\omega \sin\omega\tau, \qquad (2.80)$$

they are stroboscopic QND observables at time spaced $\tau = n\pi/\omega$. If the condition (2.78) holds at all time, it is called a continuous QND observable. For example, consider the quadratures of a harmonic oscillator,

$$\hat{X}_1(t) = \hat{x}\cos\omega t - \frac{\hat{p}}{m\omega}\sin\omega t, \qquad (2.81)$$

$$\hat{X}_2(t) = \hat{x}\cos\omega t + \frac{\hat{p}}{m\omega}\sin\omega t, \qquad (2.82)$$

which are constants of motion of a harmonic oscillator, i.e.,

$$\frac{d\hat{X}_j}{dt} = \frac{\partial \hat{X}_j}{\partial t} - \frac{i}{\hbar}\left[\hat{X}_j, \hat{H}_0\right] = 0, \tag{2.83}$$

therefore they satisfy

$$\left[\hat{X}_j(t), \hat{X}_j(t')\right] = 0 \tag{2.84}$$

for arbitrary t and t'. The quadrature $\hat{X}_1$ and $\hat{X}_2$ are continuous QND observables.

2.3.2 Requirements on the interaction

In addition to the condition on the free quantum system (2.78), in order to perform a QND measurement, one needs to carefully design the interaction between the apparatus and the system: the measurement should not perturb the observable $\hat{O}$, which means that

$$\left[\hat{O}_I(t), \hat{H}_I(t')\right] = 0, \tag{2.85}$$

where $\hat{O}_I(t)$ and $\hat{H}_I(t')$ are the interaction picture operators, i.e.,

$$\hat{O}_I(t) = \hat{U}_0^\dagger(t, t_0)\, \hat{O}(t)\, \hat{U}_0(t, t_0), \tag{2.86}$$

$$\hat{H}_I(t) = \hat{U}_0^\dagger(t, t_0)\, \hat{U}_M^\dagger(t, t_0)\, \hat{H}(t)\, \hat{U}_0(t, t_0)\, \hat{U}_M(t, t_0), \tag{2.87}$$

where $\hat{U}_0(t, t_0)$ and $\hat{U}_M(t, t_0)$ are the time evolution operators for the system and the apparatus. A simple way to satisfy this condition is to interact the observable linearly to the apparatus, i.e.,

$$\hat{H}_I = -A(t)\,\hat{O}\hat{F}, \tag{2.88}$$

where $\hat{F}$ is some observable of the apparatus and K is the coupling between the apparatus and the observable. In section 4.3.2, we will give a physical realization of QND measurement of the quadrature with an opto/electro-mechanical system.

2.4   Squeezed state of a harmonic oscillator

After discussed the effects of the Heisenberg uncertainty principle on detection (Eq. (1.9)), in this section, we will consider the consequences of the Heisenberg uncertainty principle to the quantum state of the system (Eq. (1.9)), i.e., what kind of quantum states would minimize the uncertainty relation, i.e.,

$$\Delta x \cdot \Delta p = |\langle \{\hat{x}, \hat{p}\} + [\hat{x}, \hat{p}] \rangle| /2 = \frac{\hbar}{2}. \tag{2.89}$$

The answer of this question is first given by Schrödinger in 1926 [   ], and further generalized by the others [,   ,   ,   ,   ,   ]. In order to satisfy the minimum uncertainty relation (2.89), the state needs to satisfy two conditions:

$$1. \quad \langle \psi | \{\hat{x}, \hat{p}\} | \psi \rangle = 0, \tag{2.90}$$

$$2. \quad \hat{\beta} | \psi \rangle = \beta | \psi \rangle, \tag{2.91}$$

where $\hat{\beta} = (\hat{x} + i\mu\hat{p}) / \sqrt{2\mu}$ and $\beta = (\langle \hat{x} \rangle + i\mu \langle \hat{p} \rangle) / \sqrt{2\mu}$, and condition one requires that $\mu \in \mathrm{Re}$.

Defining the annihilation operator,

$$\hat{a} = \sqrt{\frac{m\omega}{2}} \left(\hat{x} + \frac{i}{m\omega} \hat{p} \right) = \frac{1}{2} \left(\hat{X}_1 + i\hat{X}_2 \right). \tag{2.92}$$

Here we consider a harmonic oscillator with mass m and frequency ω, its motional quadratures $\hat{X}_1$ and $\hat{X}_2$ are given by

$$\hat{X}_1 = \hat{a} + \hat{a}^\dagger = \hat{x}/x_{\mathrm{zp}}, \tag{2.93}$$

$$\hat{X}_2 = -i \left(\hat{a} - \hat{a}^\dagger \right) = \hat{p}/m\omega x_{\mathrm{zp}}. \tag{2.94}$$

Now the operator $\hat{\beta}$ can be written in terms of the annihilation and creation operators as

$$\hat{\beta} = \hat{a} \cosh(r) + \hat{a}^\dagger \sinh(r) \tag{2.95}$$

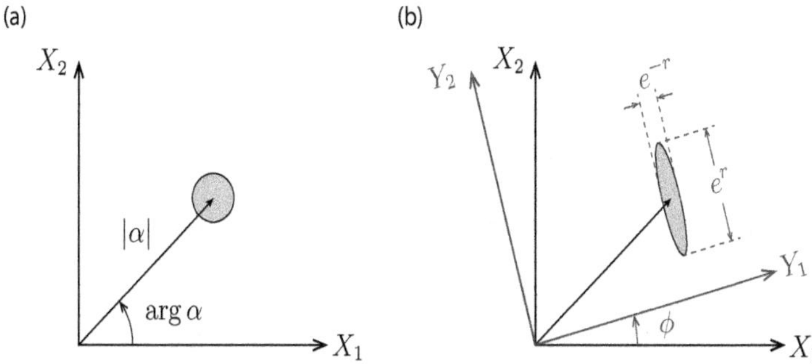

Figure 2.5: (a) Example of coherent state. (b) Example of squeezed state.

with

$$r = \cosh^{-1}\left(\frac{1 + \mu\lambda^2}{2\lambda\sqrt{\mu}}\right). \tag{2.96}$$

The operator $\hat{\beta}$ is an annihilation operator of a Bogoliubov mode, which is simply a Bogoliubov transformation of the annihilation and creation operators that conserve the commutator relation, i.e.,

$$\left[\hat{\beta}, \hat{\beta}^\dagger\right] = \left[\hat{a}, \hat{a}^\dagger\right] = 1. \tag{2.97}$$

The states that satisfy the minimum uncertainty relation can be characterized by a complex parameter $\alpha = \frac{1}{2}(X_1 + iX_2)$ and real parameter r, or $|\psi\rangle = |r, \alpha\rangle$.

2.4.1 Coherent state

First, let's consider a special case where $r = 0$. Then, condition (2.91) becomes

$$\hat{a}|\alpha\rangle = \alpha|\alpha\rangle, \tag{2.98}$$

and the quantum state that satisfy Eq. (2.98) is called coherent state, which is given by

$$|\alpha\rangle = \hat{D}(\alpha)|0\rangle = e^{-|\alpha|^2/2}\sum_{n=0}^{\infty}\frac{\alpha^n}{\sqrt{n!}}|n\rangle, \tag{2.99}$$

where $\hat{D}(\alpha) = \exp\left(\alpha\hat{a}^{\dagger} - \alpha^{*}\hat{a}\right)$ is the displacement operator. The coherent state has the following expectation values and variances:

$$\left\langle \hat{X}_1 + i\hat{X}_1 \right\rangle = 2\alpha, \tag{2.100}$$

$$\Delta X_1 = \Delta X_2 = 1, \tag{2.101}$$

$$\langle \hat{n} \rangle = |\alpha|^2, \tag{2.102}$$

$$\Delta n = \sqrt{n}, \tag{2.103}$$

where $\hat{n} = \hat{a}^{\dagger}\hat{a}$ is the number operator. Fig. 2.4a shows a coherent state in the quadrature space.

2.4.2 Squeezed state

Having obtained the special solution for $r = 0$, finding the eigenstates of the Bogoliubov mode $\hat{\beta}$ is equivalent to finding an operator $\hat{S}$ which can transform the Bogoliubov mode $\hat{\beta}$ into $\hat{a}$ by means of similarity transformation, which is given by the unitary squeeze operator [,]

$$\hat{S}[\epsilon] = \exp\left[\frac{1}{2}\left(\epsilon^{*}\hat{a}^2 - \epsilon\hat{a}^{\dagger 2}\right)\right]. \tag{2.104}$$

Here we consider a more general transformation that $\epsilon = re^{2i\phi}$, which is a complex number. The squeeze operator obeys the relations

$$\hat{S}^{\dagger}(\epsilon) = \hat{S}^{-1}(\epsilon) = \hat{S}(-\epsilon) \tag{2.105}$$

and has the following transformation properties:

$$\hat{S}(\epsilon)\,\hat{a}\hat{S}^{\dagger}(\epsilon) = \hat{a}\cosh r + \hat{a}^{\dagger}e^{-2i\phi}\sinh r, \tag{2.106}$$

$$\hat{S}(\epsilon)\,\hat{a}^{\dagger}\hat{S}^{\dagger}(\epsilon) = \hat{a}^{\dagger}\cosh r + \hat{a}e^{-2i\phi}\sinh r, \tag{2.107}$$

$$\hat{S}(\epsilon)\left(\hat{Y}_1 + i\hat{Y}_2\right)\hat{S}^{\dagger}(\epsilon) = \hat{Y}_1 e^{r} + i\hat{Y}_2 e^{-r}, \tag{2.108}$$

where

$$\hat{Y}_1 + i\hat{Y}_1 = \left(\hat{X}_1 + i\hat{X}_2\right) e^{-i\phi} \tag{2.109}$$

is the rotated complex amplitude. When $\phi = 0$, then Eq. (2.106) becomes (2.95). The eigenstate of the Bogoliubov operator $\hat{\beta}$ is given by applying the squeeze operator $\hat{S}(\epsilon)$ to the coherent state $|\alpha\rangle$, which is

$$|\alpha, \epsilon\rangle = \hat{D}(\alpha)\,\hat{S}(\epsilon)\,|0\rangle, \tag{2.110}$$

this state is called squeezed state, which has the following expectation values

$$\left\langle \hat{X}_1 + i\hat{X}_2 \right\rangle = \left\langle \hat{Y}_1 + i\hat{Y}_2 \right\rangle e^{i\phi} = 2\alpha, \tag{2.111}$$

$$\Delta Y_1 = e^r, \quad \Delta Y_2 = e^r, \tag{2.112}$$

$$\langle \hat{n} \rangle = |\alpha|^2 + \sinh^2 r, \tag{2.113}$$

$$(\Delta n)^2 = \left| \alpha \cosh r - \alpha^* e^{2i\phi} \sinh r \right|^2 + 2\cosh^2 r \sinh^2 r. \tag{2.114}$$

The squeezed state has unequal uncertainties in the quadratures Y_1 and Y_2, as shown by the error ellipse in in Fig. 2.5. The principle axes of the ellipse lie along the rotated quadratures Y_1 and Y_2. Note that

$$\Delta x \cdot \Delta p = \frac{\hbar}{2}\Delta X_1 \Delta X_2 = \frac{\hbar}{2}\sqrt{1 + 4\sinh^2 r \cosh^2 r \sin^2 2\phi}, \tag{2.115}$$

only when $\phi = 0$, the squeezed state minimizes the uncertainty relation (2.89).

2.4.3 Discussion

The minimum uncertainty state is one of the most important type of quantum states, which has long been applied to various fields of theoretical physics such as quantum field theory, quantum statistical mechanics, solid state physics, etc. [,   ,   ]. These quantum states become widely recognized in the experimental side (particularly in optics) during the 1960's due to the works of Glauber [ ], Klauder [,   ], and Sudarshan [ ]. Together with the invention of maser and laser, which makes these quantum states become available in experiment. These states provide a lot of applications, such as ultra-sensitive detection [ ] and quantum information processing [,].

The statistical properties of the coherent light were first observed by Arechi in 1965 [ ]. Twenty years later, squeezed light was observed by several groups in various systems [, , ,]. Nowadays, More than 10 dB squeezing has been realized via optical parametric oscillation [ ]. Recently, squeezed light has been generated in an on-chip optomechanical system [ ]. Besides electromagnetic wave, the quantum squeezed states of motion have been observed in microscopic degrees of freedom such as the motion of a trapped ion [,] or lattice vibration in solid [ ]. In macroscopic degrees of freedom such as the center of mass motion of a nanomechanical resonator, thermal squeezed state was first demonstrated by Rugar et al. with the parametric driving technique [ ]. Until recently, quantum squeezed states of nanomechanical resonators have been generated [, , ,] via the reservoir engineering squeezing technique [ ] in electromechanical systems. We will introduce this dissipative squeezing technique in section 4.3.3.

Chapter 3

Cavity electromechanics with superconducting circuits

This chapter describes the procedure of building a cavity electromechanical system out of electrical circuits and a mechanical oscillator. First, a classical description of circuits and the mechanics will be given. Because the experimental temperature is 10 mK and the energy scales of the circuit ($\sim$ 5 GHz) correspond to temperature of 250 mK, a quantum description of the circuits and the mechanical oscillator is used to fully understand the behavior of the electromechanical system. Connection between the language of microwave engineering to that of quantum optics will be discussed. At the end, the quantum theory is used to describe the cavity electromechanical system.

3.1   Lumped element microwave resonator

In cavity opto/electromechanics, a harmonic mode in the cavity is coupled to the mechanical oscillator. In microwave domain, the cavity can be realized using electrical circuits. Fig. 3.1a is the optical micrograph of our device, which is a lumped element superconducting microwave resonator: a parallel plate vacuum gap capacitor in parallel with a spiral inductor. The mechanical element in this device is the top gate of the capacitor, which is a high Q suspended aluminum membrane. In order to excite and read-out the microwave resonance, the LC resonator is capacitively coupled to transmission lines. The equivalent circuit is shown in 3.1b, the lumped element resonator is represented by the RLC resonator (blue), and the couplers are represented by the capacitors on the input and the output sides

(a) (b) (c)

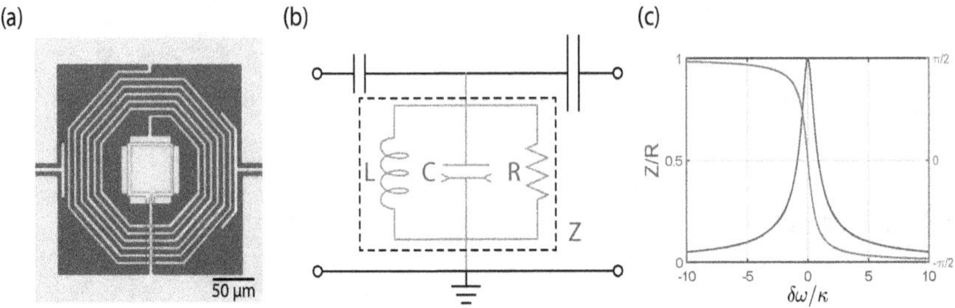

Figure 3.1: Cavity electromechanics. (a) An optical micrograph of a circuit cavity electromechanical system, which is a lumped element LC resonator with a movable capacitor. (b) The equivalent circuit of the circuit cavity electromechanical system in (a). (c) Impedance of the RLC resonator, the blue curve is the amplitude and the red curve is the phase.

(black).

In this section, we will focus on the microwave resonator (blue lines in Fig. 3.1b). By applying the Kirchhoff's current law, the equation of motion of the magnetic flux in the inductor is

$$\frac{d^2\Phi}{dt^2} + \frac{1}{RC}\frac{d\Phi}{dt} + \frac{\Phi}{LC} = 0, \tag{3.1}$$

and the solution of this differential equation is

$$\Phi(t) = \Phi_0 \exp\left[i\left(\omega_c + i\kappa\right)t + \phi\right], \tag{3.2}$$

where the frequency $\omega_c = 1/\sqrt{LC}$ and the decay rate $\kappa = 1/RC$. In frequency domain, the circuit can be described by its impedance $Z(\omega)$, which is given by

$$Z(\omega) = \left(j\omega C + \frac{1}{j\omega L} + \frac{1}{R}\right)^{-1}. \tag{3.3}$$

Near resonance, we can expand $Z(\omega)$ to first order in $\delta\omega = \omega - \omega_c$, which yields

$$Z(\omega) \simeq \frac{R}{1 + 2jQ\delta\omega/\omega_c}, \tag{3.4}$$

where $Q = \omega_c/\kappa$ is the quality factor of the cavity, which is the number of oscillation within the cavity decay time $\tau = 1/\kappa = RC$ before the system comes to equilibrium. Fig. 3.1c

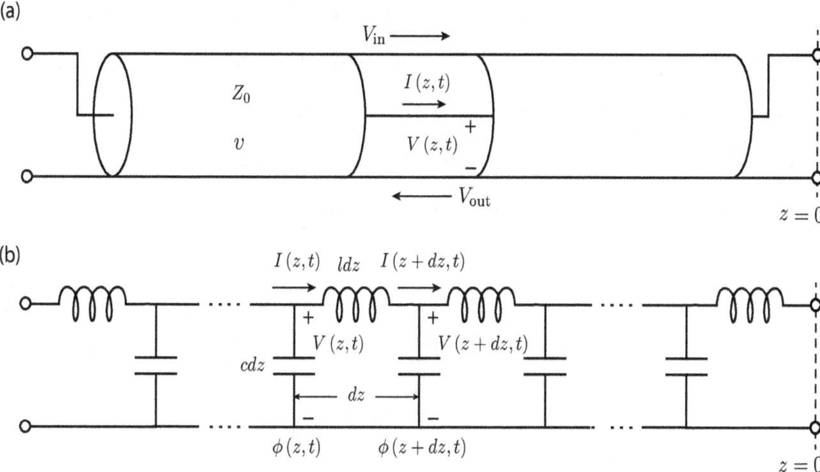

Figure 3.2: Transmission line. (a) A schematic of a transmission line with impedance Z_0 and phase velocity v. $V_{\text{in,out}}$ are the incoming and outgoing voltages (relative to $z = 0$). $I(z,t)$ and $V(z,t)$ are the current and voltage. (b) The equivalent circuit of the transmission line in (a). c and l are the capacitance and inductance per unit length, $\phi(z,t)$ is the node phase.

shows the real and imaginary part of the impedance $Z(\omega)$. The motion of the mechanical oscillator varies the resonance frequency of the LC circuit through the modulation of the capacitance. By measuring the frequency modulation of the impedance, we can detect the mechanical motion.

3.2   Classical transmission line theory

In order to probe the resonator, we need to connect the circuit to the outside world. In microwave domain, it can be realized by connecting the electrical resonator to transmission line. In this section, we follow [ ] to give an introduction to the transmission line theory and the classical input-output theory.

3.2.1   Transmission line theory

Fig. 3.2a is the coaxial transmission line, which is generally used to guide microwave signals. Because the physical dimensional of the transmission line is much longer than the wavelength of the microwave signal used to drive the circuit, the voltages and currents can vary in magnitude and phase along the line. An ideal transmission line can be considered as a perfectly conducting wire with inductance per unit length of l and capacitance to ground per unit length c, the equivalence circuit of the transmission line is given by Fig. 3.2b.

Applying the Kirchhoff's voltage and current law, we arrive at the dynamical equations of the voltages and currents in the transmission line,

$$\frac{\partial V}{\partial z}(z,t) = -l\frac{\partial I}{\partial t}(z,t), \tag{3.5}$$

$$\frac{\partial I}{\partial z}(z,t) = -c\frac{\partial V}{\partial t}(z,t). \tag{3.6}$$

Eqs (3.5) and (3.6) can be decoupled by introducing the left and right propagating modes

$$V(z,t) = V_R(z,t) + V_L(z,t), \tag{3.7}$$

$$I(z,t) = \frac{1}{Z_0}[V_R(z,t) - V_L(z,t)], \tag{3.8}$$

where $Z_0 = \sqrt{l/c}$ is the characteristic impedance of the transmission line, which gives

$$\frac{\partial V_R}{\partial t} + v\frac{\partial V_R}{\partial x} = 0, \tag{3.9}$$

$$\frac{\partial V_L}{\partial t} - v\frac{\partial V_L}{\partial x} = 0, \tag{3.10}$$

where $v = 1/\sqrt{lc}$ is the phase velocity of the electromagnetic wave in the transmission line. Since the transmission line is dispersionless (v is constant), the solution of these equations can be propagated by uniform translation without changing shape,

$$V_R(z,t) = V_{\text{in}}\left(t - \frac{z}{v}\right), \tag{3.11}$$

$$V_L(z,t) = V_{\text{out}}\left(t + \frac{z}{v}\right), \tag{3.12}$$

where V_{in} and V_{out} are arbitrary functions. Here, we define the right propagating mode as the incoming mode and the left propagating mode as the outgoing mode. For an infinite transmission line, the right propagating mode V_{in} and the left propagating mode V_{out} are

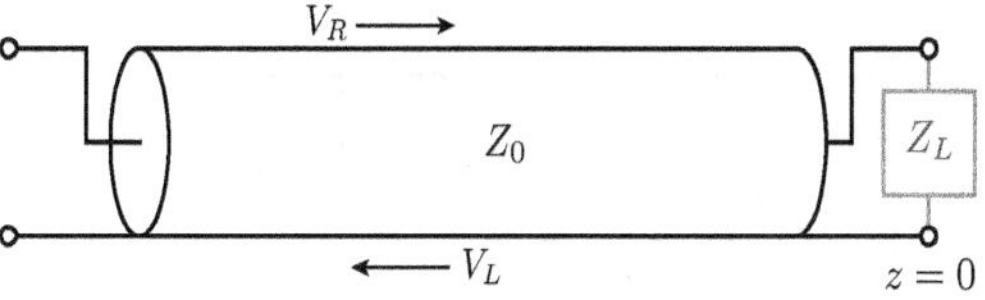

Figure 3.3: A transmission loaded by a system with impedance Z.

completely independent.

For most of the applications, the transmission line is terminated at $z = 0$ by some system (Fig. 3.3a). In this case, the two solutions (3.12) and (3.11) will be related by the boundary condition,

$$V(z = 0, t) = V_{\text{in}}(t) + V_{\text{out}}(t), \tag{3.13}$$

$$I(z = 0, t) = \frac{V_{\text{in}}(t)}{Z_0} - \frac{V_{\text{out}}(t)}{Z_0}. \tag{3.14}$$

We can rearrange Eq. (3.14) to

$$V_{\text{out}}(t) = V_{\text{in}}(t) - Z_0 I(z = 0, t). \tag{3.15}$$

The first term in the right hand side is the contribution of the incoming wave, the second term is from the current injected by the system dynamics. Therefore, one can probe the system by measuring the output field of a transmission line. The net power flowing into the load is equal to

$$P = \frac{1}{Z}\left[V_{\text{in}}^2(t) - V_{\text{out}}^2(t)\right]. \tag{3.16}$$

3.2.2 The effects of a transmission line to the system

The transmission line not only enables us to probe the system, but also loads the system and influence its dynamics. For example, if we turn off the input field in Eqs. (3.13) and (3.14), the response of the transmission line at $z = 0$ is equal to the response of a resistance with resistivity equal to the characteristic impedance of the line Z_0. The semi-infinite

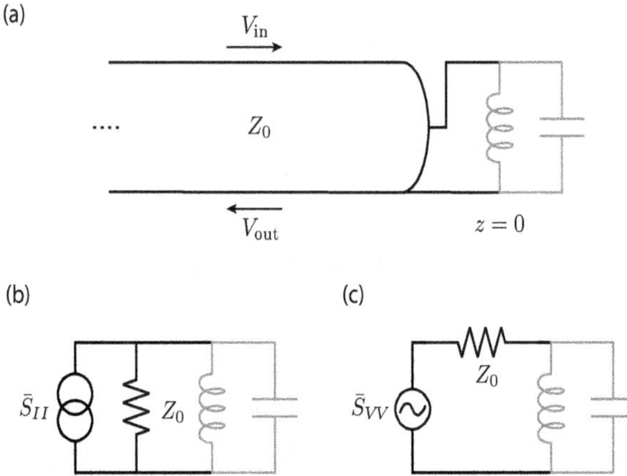

Figure 3.4: (a) A transmission line loaded by an LC circuit. (b)(c) The equivalent circuit of (a), where the effect of the transmission line to the LC circuit is equivalent to a noisy resistor. (b) Noise current source in parallel to a resistor. (c) Noise voltage source in series with a resistor.

transmission line simply carries the energy away from the system as propagating wave. The infinite extent of the line allow us to introduce dissipation with only reactive elements, which enables us to formulate a quantum theory of a resistor, we will explore this topic later.

Although the transmission line induces dissipation, it also enables us to control the system. To make it clear, let's combine Eqs. (3.13) and (3.14) to eliminate the output field V_{out}, then we can express the system current and voltage in terms of the input field,

$$I(z = 0, t) = \frac{2V_{\text{in}}(t)}{Z_0} - \frac{V(z = 0, t)}{Z_0}, \tag{3.17}$$

where the first term in the right hand side is the external drive that is used to control the system, the second term describes the dissipation induced by the transmission line. For example, consider a LC resonator couples to a semi-infinite transmission line as shown in Fig. 3.4a. Using the Kirchhoff' current law, the equation of motion of the magnetic flux in the inductor is

$$\frac{d^2\Phi}{dt^2} + \frac{1}{Z_0 C}\frac{d\Phi}{dt} + \frac{\Phi}{LC} = \frac{2}{Z_0 C}V_{\text{in}}(t), \tag{3.18}$$

the second term in Eq. (3.17) results in the damping.

3.2.3   Classical statistical mechanics of a transmission line

Compare Eq. (3.18) to Eq. (3.1), the transmission line behaves effectively as a resistor. The effect of the transmission line to the LC resonator is equivalent to a thermal bath. Consider connecting a LC resonator to a transmission line at temperature T without external drive, the thermal fluctuation of the transmission line would drive the system into thermal equilibrium. In this case, Eq. (3.18) is the Langevin equation and the input voltage $V_{\text{in}}(t)$ represents the Langevin force from the bath (the transmission line). To obtain the spectral density of the fluctuating field $V_{\text{in}}(t)$, one can write the solution of Eq. (3.18) in frequency space and solve for $\Phi[\omega]$, which gives

$$\Phi[\omega] = K[\omega]I_{\text{in}}[\omega] = \frac{2/Z_0 C}{(\omega_c^2 - \omega^2) - i\omega/Z_0 C}V_{\text{in}}[\omega], \tag{3.19}$$

where $K[\omega]$ is the response of the system. The spectrum of the flux is related to the spectrum of the fluctuating input voltage by

$$\bar{S}_{\Phi\Phi}[\omega] = |K[\omega]|^2 \bar{S}_{V_{\text{in}}V_{\text{in}}}[\omega]. \tag{3.20}$$

Here $\bar{S}[\omega] = S[\omega] + S[-\omega]$ is the single side spectral density. In thermal equilibrium, the energy of the LC resonator is

$$\bar{E} = 2 \times \frac{1}{2L}\langle \Phi^2 \rangle = \frac{1}{L}\int_0^\infty \frac{d\omega}{2\pi}\bar{S}_{\Phi\Phi}[\omega] = k_B T. \tag{3.21}$$

For high Q resonator ($\omega_c \gg 1/Z_0 C$), the response $|K[\omega]|^2$ is a sharp resonance at the resonance freqeuncy ω_c, and therefore

$$\bar{E} = \bar{S}_{V_{\text{in}}V_{\text{in}}}[\omega_c]\frac{1}{L}\int_0^\infty \frac{d\omega}{2\pi}|K[\omega]|^2 = \bar{S}_{V_{\text{in}}V_{\text{in}}}[\omega_c]/Z_0 = k_B T. \tag{3.22}$$

In principle, one can couple LC resonators at arbitrary resonance frequency without changing the interaction with the heat bath. Therefore, Eq. (3.22) implies

$$\bar{S}_{V_{\text{in}}V_{\text{in}}}[\omega] = Z_0 k_B T. \tag{3.23}$$

A transmission line at temperature T generates a white input voltage noise. This relation between the spectral density of the Langevin force and the dissipation induced by the bath is the famous fluctuation-dissipation theorm (or Nyquist's theorem).

If we define a noise current $I_n = \frac{2V_{\text{in}}}{Z_0}$, the transmission line is equivalent to a noisy current source in parallel to a resistor (Fig. 3.4b) with current noise

$$\bar{S}_{I_n I_n}[\omega] = \frac{4k_B T}{Z_0}. \tag{3.24}$$

It can also be described by it's Noton's equivalent circuit (Fig. 3.4b), which is a noisy voltage source ($V_n = Z_0 I_n$) in series with a resistor with voltage noise

$$\bar{S}_{V_n V_n}[\omega] = Z_0^2 \bar{S}_{I_n I_n}[\omega] = 4Z_0 k_B T. \tag{3.25}$$

3.2.4 Capacitively coupled RLC resonator

In the previous sections, we have discussed a LC resonator directly couples to a transmission line. The quality factor of a resonator couple to a transmission line with impedance Z_0 is

$$Q = \frac{\omega_c}{\kappa} = \omega_c Z_0 C. \tag{3.26}$$

For the lumped element resonator in our experiment, the resonance frequency $\omega_c \simeq 2\pi \times$ 5GHz and the capacitance $C \simeq 100\text{fF}$. If we directly connect it to a 50Ω transmission line, the external quality factor would be too small ($Q_{\text{ext}} \simeq 0.1$). One way to reduce the dissipation from the transmission line is to couple it through a small capacitor [ ].

In our experiment, two transmission lines are capacitively coupled to the resonator with input coupling capacitor C_L and output coupling capacitor C_R (Fig. 3.5a), which has the equivalent circuit Fig. 3.5b. Here we account for the internal losses of the resonator with

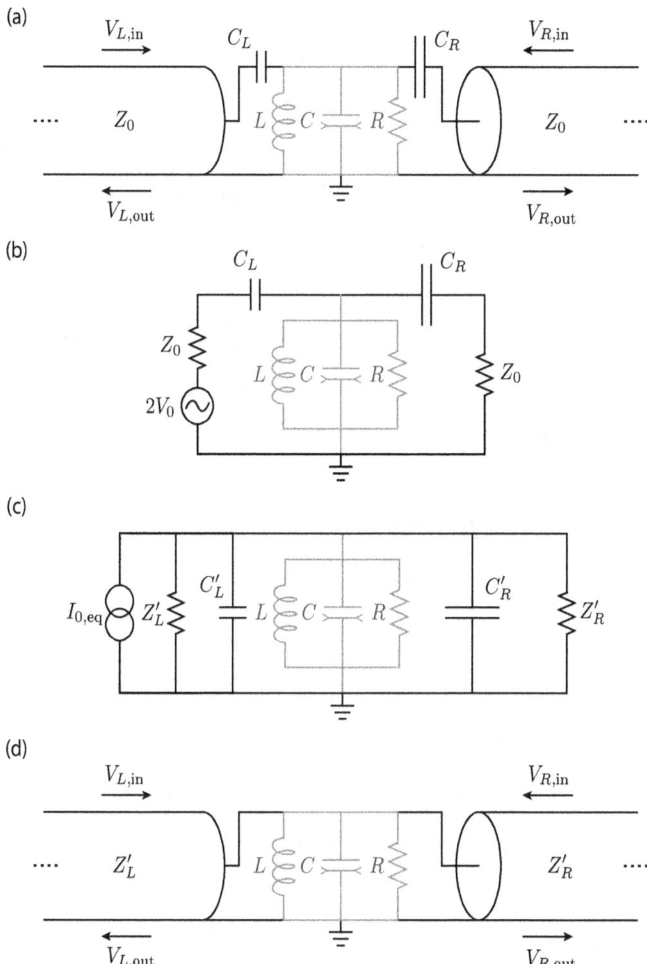

Figure 3.5: (a) An RLC resonator capacitively coupled to two transmission lines with coupling capacitors $C_{L,R}$. (b) Equivalent circuit of (a). (c) The Norton equivalent circuit of (b). The series coupling capacitors $C_{L,R}$ and impedances Z_0 have been replaced with the Norton equivalent parallel network with capacitors $C'_{L,R}$ and impedances Z'_0. (d) Capacitively coupled transmission line with impedance Z_0 and capacitor C_σ is equivalent to directly coupled transmission line with impedance Z'_σ.

68

the resistor R. The Norton equivalent circuit is shown in Fig. 3.5c with

$$Z'_\sigma = \frac{1 + q_\sigma^2}{q_\sigma^2} Z_0, \tag{3.27}$$

$$C'_\sigma = \frac{1}{1 + q_\sigma^2} C_\sigma, \tag{3.28}$$

where $q_\sigma = \omega Z_0 C_\sigma$, $\sigma = L, R$. For our device, $C_\sigma \simeq 1$ fF and $q_\sigma \simeq 1.6 \times 10^{-3} \ll 1$, therefore $Z'_\sigma \simeq Z_0/q_\sigma^2$ and $C'_\sigma \simeq C_\sigma$. The total capacitance becomes $C_{\text{tot}} = C + C_L + C_R$. Since $C \gg C_{R,L}$ for our device, we will neglect the resonance frequency shift due to the coupling capacitances ($\omega'_c = 1/\sqrt{LC_{\text{tot}}} \simeq \omega_c$). Therefore, couples a transmission line with impedance Z_0 and coupling capacitance C_σ is equivalent to directly couples a transmission line with impedance Z'_σ (Fig. 3.5d). The quality factor of the loaded resonator is given by $Q = (\frac{1}{Q_{\text{int}}} + \frac{1}{Q_L} + \frac{1}{Q_R})^{-1}$ with

$$Q_{\text{int}} = \frac{R}{Z_{LC}}, \tag{3.29}$$

$$Q_{L,R} = \frac{Z'_0}{Z_{LC}} = \frac{1}{q_{L,R}^2} \frac{Z_0}{Z_{LC}}, \tag{3.30}$$

where $Z_{LC} = \sqrt{L/C_{\text{tot}}}$. The external quality factor $Q_{L,R}$ is enhanced by the factor $1/q_{L,R}^2$. For our device, $Q_{L,R}$ is around 40×10^3.

For a RLC resonator couples to two transmission lines as shown in Fig. 3.5d, the equation of motion of the magnetic flux Φ in the inductor can be obtained by generalizing Eq. (3.17) to

$$I(z = 0, t) = \sum_{\sigma=I,L,R} \left(\frac{2V_{\sigma,\text{in}}(t)}{Z'_\sigma} - \frac{V(z = 0, t)}{Z'_\sigma} \right), \tag{3.31}$$

where I represent the internal port with $Z'_I = R$ that account for the dissipation from the resistor, and L, R represent the input and output transmission lines with $Z'_{L,R} = Z_0/q_{L,R}^2$. Then, the equation of motion of the magnetic flux in the inductor is

$$\frac{d^2\Phi}{dt^2} + \kappa \frac{d\Phi}{dt} + \omega_c^2 \Phi = 2\kappa V_{\text{in}}(t), \tag{3.32}$$

where

$$\kappa = \sum_{\sigma=I,L,R} \kappa_\sigma, \tag{3.33}$$

$$V_{\text{in}}(t) = \sum_{\sigma=I,L,R} \frac{\kappa_\sigma}{\kappa} V_{\sigma,\text{in}}(t), \tag{3.34}$$

$$\kappa_\sigma = 1/Z'_\sigma C_{\text{tot}}. \tag{3.35}$$

If the resistor and the transmission lines are in thermal equilibrium at temperature T, according to (3.23) the noise spectrum of $V_{\sigma,\text{in}}$ is

$$\bar{S}_{V_{\sigma,\text{in}}V_{\sigma',\text{in}}}[\omega] = Z_\sigma k_B T \delta_{\sigma,\sigma'}. \tag{3.36}$$

3.2.5 Classical Langevin equation of a mechanical oscillator

Although the discussion in the previous sections focus on the dynamics of a LC circuit couple to a transmission line, the results can be applied to describe a mechanical oscillator coupled to a dissipative environment by making the following analogue in Eq. (3.18):

$$\begin{aligned}
C &\rightarrow m, \\
\frac{1}{Z_0} &\rightarrow m\Gamma_m, \\
\frac{1}{LC} &\rightarrow \omega_m^2, \\
\Phi &\rightarrow x, \\
\frac{2}{Z_0} V_{\text{in}} &\rightarrow F_n,
\end{aligned} \tag{3.37}$$

where m is the mass of the oscillator, Γ_m is the damping rate, and ω_m is the mechanical resonance frequency, which gives the Langevin equation of a mechanical oscillator

$$m\frac{d^2x(t)}{dt^2} + m\Gamma_m \frac{dx(t)}{dt} + m\omega_m^2 x = F_n(t). \tag{3.38}$$

The force noise of the bath is given by

$$\bar{S}_{F_n F_n}[\omega] = 4m\Gamma_m k_B T. \tag{3.39}$$

3.3 Classical electromechanics in dissipative environment

After introducing the Langevin equations of the LC circuit and the mechanical oscillator, we are now in the position to describe the classical dynamics of an electromechanical system.

The capacitance of the LC circuit is modulated by the motion of the membrane. For small displacement from the equilibrium position, the capacitance is linearly related to the displacement

$$C_{\text{tot}}(x) = C_{\text{tot}}(0) + \frac{\partial C}{\partial x}x. \tag{3.40}$$

The coupling between the LC circuit and the motion of the membrane is given by the capacitive energy

$$E_{\text{cap}} = \frac{1}{2}C_{\text{tot}}(x)V^2, \tag{3.41}$$

which gives the radiation pressure

$$F_{\text{rad}} = -\frac{\partial E_{\text{cap}}}{\partial x} = -\frac{1}{2}\frac{\partial C}{\partial x}V^2. \tag{3.42}$$

The classical dynamics of a driven electromechanical system is governed by the nonlinear coupled Langevin equations

$$C_{\text{tot}}(x)\frac{d^2\Phi}{dt^2} + \frac{1}{Z_0}\frac{d\Phi}{dt} + \frac{\Phi}{L} = I_n(t) + I_d(t), \tag{3.43}$$

$$m\frac{d^2x}{dt^2} + m\Gamma_m\frac{dx}{dt} + m\omega_m^2 x = F_n(t) - \frac{1}{2}\frac{\partial C}{\partial x}[V(t)]^2, \tag{3.44}$$

where $I_n(t)$ and $F_n(t)$ are the noise current and the Langevin force from the dissipative environments, and $I_d(t)$ is the external drive current. The output voltage of the transmission line is given by

$$V_{\sigma,\text{out}}(t) = V_{\sigma,\text{in}}(t) - Z_\sigma I(t), \tag{3.45}$$

with

$$I(t) = \sum_{\sigma=I,L,R} \left(\frac{2V_{\sigma,\text{in}}(t)}{Z_\sigma} - \frac{V(t)}{Z_\sigma} \right).$$ (3.46)

In general case, the nonlinear Langevin equations (3.43) and (3.44) have no analytical solution without making approximation. For high Q microwave resonator and mechanical oscillator in strong driving regime as considered in this work, several approximations can be made to simplify these equations. The solutions of the classical equations have been discussed in [ ]. We will postpone the discussion of the solutions after quantizing the system.

3.4 Quantization of a harmonic oscillator and a transmission line

This section briefly discusses the procedure to quantize a harmonic oscillator, which can be an electrical circuit or mechanical oscillator. The results are extended to quantize a transmission line.

3.4.1 Quantization of a harmonic oscillator

Let's begin with a harmonic oscillator with mass m and frequency ω_m. The Lagrangian of a harmonic oscillator is

$$L = \frac{1}{2}m\dot{x}^2 - \frac{1}{2}m\omega_m^2 x^2.$$ (3.47)

The corresponding canonical conjugate momentum is

$$p = \frac{\partial L}{\partial \dot{x}} = m\dot{x},$$ (3.48)

from which we obtain the Hamiltonian

$$H = p\dot{x} - L = \frac{p^2}{2m} + \frac{1}{2}m\omega_m^2 x^2.$$ (3.49)

To quantize the harmonic oscillator, we replace the classical variable x and p by the corresponding operators $\hat{x}$ and $\hat{p}$, and impose the canonical commutation relation

$$[\hat{x}, \hat{p}] = i\hbar. \tag{3.50}$$

Define the annihilation and creation operators

$$\hat{b} = \sqrt{\frac{m\omega_m}{2\hbar}} \left(\hat{x} + i\frac{\hat{p}}{m\omega_m} \right), \tag{3.51}$$

$$\hat{b}^\dagger = \sqrt{\frac{m\omega_m}{2\hbar}} \left(\hat{x} - i\frac{\hat{p}}{m\omega_m} \right), \tag{3.52}$$

with the commutation relation

$$\left[\hat{b}, \hat{b}^\dagger \right] = 1. \tag{3.53}$$

The position and momentum operators can be expressed in terms of the creation and annihilation operators as

$$\hat{x} = x_{zp} \left(\hat{b} + \hat{b}^\dagger \right), \tag{3.54}$$

$$\hat{p} = -m\omega_m x_{zp} i \left(\hat{b} - \hat{b}^\dagger \right), \tag{3.55}$$

where $x_{zp} = \sqrt{\hbar/2m\omega_m}$ is the zero-point motion. The Hamiltonian (3.48) can be written as

$$\hat{H} = \hbar\omega_m \left(\hat{N} + \frac{1}{2} \right), \tag{3.56}$$

where $\hat{N} = \hat{b}^\dagger \hat{b}$ is the number operator with discrete eigenvalue $n = 0, 1, 2, ...$ Therefore, a quantum harmonic oscillator has discrete energy $E_n = \hbar\omega_m(n + \frac{1}{2})$. Note that even at the ground state ($n = 0$), a quantum harmonic oscillator has non-zero $E_0 = \hbar\omega_m/2$. This minimum energy is the zero-point energy, which is the consequence of the wave-like nature of matter.

The same procedures can be applied to quantize an LC circuit. The Lagrangian of a LC is

$$L = \frac{1}{2}C\dot{\Phi}^2 - \frac{1}{2L}\Phi^2, \tag{3.57}$$

where Φ is the magnetic flux in the inductor. The corresponding conjugate canonical momentum is

$$Q = \frac{\partial L}{\partial \dot{\Phi}} = C\dot{\Phi}, \tag{3.58}$$

which is the charge in the capacitor. The Hamiltonian is given by

$$H = Q\dot{\Phi} - L = \frac{Q^2}{2C} + \frac{1}{2L}\Phi^2. \tag{3.59}$$

Compare (3.57) to (3.49), we can transfer the results of the harmonic oscillator to the quantized LC circuit simply by changing the notations:

$$\begin{aligned} m &\to C, \\ \omega_m &\to \omega_c = 1/\sqrt{LC}, \\ x &\to \Phi, \\ p &\to Q. \end{aligned} \tag{3.60}$$

Then we have

$$\left[\hat{\Phi}, \hat{Q}\right] = i\hbar. \tag{3.61}$$

$$\hat{\Phi} = \sqrt{\frac{\hbar Z}{2}}\left(\hat{a} + \hat{a}^\dagger\right), \tag{3.62}$$

$$\hat{Q} = -i\sqrt{\frac{\hbar}{2Z}}\left(\hat{a} - \hat{a}^\dagger\right), \tag{3.63}$$

$$\hat{H} = \hbar\omega_c\left(\hat{N} + \frac{1}{2}\right), \tag{3.64}$$

where $\hat{a}(\hat{a}^\dagger)$ is the annihilation (creation) operator of the resonator mode, $Z = \sqrt{\frac{L}{C}}$ is the characteristic impedance of a LC circuit.

3.4.2 Quantization of a transmission line

In the following, we will extend the results of the quantized LC circuit to quantize a transmission line [ ]. As shown in Fig. 3.2, a transmission line with impedance Z_0 and phase velocity v can be described by a chain of LC resonator with capacitance per unit length c and inductance per unit length l. The Lagrangian of a transmission line of length d is

$$L = \int_0^d \left[\frac{1}{2} c \dot{\phi} (z,t)^2 - \frac{1}{2l} \left(\frac{\partial \phi}{\partial z} (z,t) \right)^2 \right] dz, \tag{3.65}$$

where $\phi(z,t)$ is the node flux. The equation of motion of $\phi(z,t)$ is given by the Eular-Lagrange equation $\frac{\partial \mathcal{L}}{\partial \phi} - \partial_i \left(\frac{\partial \mathcal{L}}{\partial (\partial_i \phi)} \right)$,

$$\frac{\partial^2 \phi}{\partial t^2} - v^2 \frac{\partial^2 \phi}{\partial z^2} = 0, \tag{3.66}$$

where $v = \frac{1}{\sqrt{lc}}$ is the phase velocity of the transmission line. A general solution can be written as

$$\phi(x,t) = \sum_{n=1}^{\infty} \Phi_n(t) \cos(k_n z + \alpha_n), \tag{3.67}$$

with

$$\Phi_n(t) = A_n \cos(\omega_n t + \beta_n), \tag{3.68}$$

where k_n, α_n, A_n, and β_n are depended on the boundary conditions. The frequency $\omega_n = v k_n$.

For an open transmission line, the boundary condition is $I = \frac{1}{l} \frac{\partial \phi}{\partial x}|_{z=0,d} = 0$, which gives $\alpha_n = 0$ and $k_n = n\pi/d$. If we substitute the solution (3.67) into (3.65), the Lagrangian of the transmission becomes

$$L = \sum_{n=1}^{\infty} \left(\frac{1}{2} C_n \dot{\Phi}_n^2 - \frac{1}{2 L_n} \Phi_n^2 \right), \tag{3.69}$$

which is equivalent to the Lagrangian of infinite uncoupled LC resonators with capacitances $C_n = cd/2$ and inductances $L_n = \frac{2ld}{(n\pi)^2}$. The canonical conjugate momentum is given by

$$Q_n = \frac{\partial L}{\partial \dot{\Phi}_n} = C \dot{\Phi}_n, \tag{3.70}$$

and the Hamiltonian is

$$H = \sum_{n=1}^{\infty} Q_n \dot{\Phi}_n - L = \sum_{n=1}^{\infty} \left(\frac{Q_n^2}{2C_n} + \frac{1}{2L_n} \Phi_n^2 \right). \tag{3.71}$$

We can follow the procedures in the last section to quantize the individual LC resonator in the transmission line.

We first impose the canonical commutation relation

$$\left[\hat{\Phi}_n, \hat{Q}_m \right] = i\hbar \delta_{n,m}. \tag{3.72}$$

The magnetic flux Φ_n and the charge Q_n can be written as

$$\hat{\Phi}_n = \sqrt{\frac{\hbar Z_n}{2}} \left(\hat{a}_n + \hat{a}_n^{\dagger} \right), \tag{3.73}$$

$$\hat{Q}_n = -i\sqrt{\frac{\hbar}{2Z_n}} \left(\hat{a}_n - \hat{a}_n^{\dagger} \right), \tag{3.74}$$

where $Z_n = \sqrt{L_n/C_n} = 2Z_0/n\pi$, $\hat{a}_n$ and $\hat{a}_n^{\dagger}$ are the annihilation and creation operators of the mode in the n th LC resonator in the transmission line, and they satisfy the commutation relation

$$\left[\hat{a}_n, \hat{a}_m^{\dagger} \right] = \delta_{n,m}. \tag{3.75}$$

Substitute (3.73) and (3.74) into (3.71), the Hamiltonian becomes

$$\hat{H} = \sum_{n=1}^{\infty} \hbar \omega_n \left(\hat{a}_n^{\dagger} \hat{a}_n + \frac{1}{2} \right), \tag{3.76}$$

where $\omega_n = 1/\sqrt{L_n C_n}$.

After quantized the transmission line. The node flux operator corresponding to (3.67) is given by

$$\hat{\phi}(z,t) = \sum_{n=1}^{\infty} \sqrt{\frac{\hbar Z_n}{2}} \left(\hat{a}_n e^{-i\omega_n t} + \hat{a}_n^{\dagger} e^{i\omega_n t} \right) \cos(k_n z), \tag{3.77}$$

and the voltage is

$$\hat{V}(z,t) = \frac{\partial \phi}{\partial t}(z,t) = -i\sqrt{\frac{2v}{d}} \sum_{n=1}^{\infty} \sqrt{\frac{\hbar \omega_n Z_0}{2}} \left(\hat{a}_n e^{-i\omega_n t} - \hat{a}_n^\dagger e^{i\omega_n t} \right) \cos(k_n z). \tag{3.78}$$

The voltage (3.78) can be written as the sum of the right propagating solution $\hat{V}_{\text{in}}$ and left propagating solutions $\hat{V}_{\text{out}}$

$$\hat{V}(z,t) = \hat{V}_{\text{in}}\left(t - \frac{z}{v}\right) + \hat{V}_{\text{out}}\left(t + \frac{z}{v}\right), \tag{3.79}$$

with

$$\hat{V}_{\text{in}}\left(t - \frac{z}{v}\right) = -i \int_0^\infty \frac{d\omega}{2\pi} \sqrt{\frac{\hbar \omega Z_0}{2}} \left\{ \hat{a}_{\text{in}}[\omega] e^{-i\omega(t-\frac{z}{v})} - (\hat{a}_{\text{in}}[\omega])^\dagger e^{i\omega(t-\frac{z}{v})} \right\}, \tag{3.80}$$

$$\hat{V}_{\text{out}}\left(t + \frac{z}{v}\right) = -i \int_0^\infty \frac{d\omega}{2\pi} \sqrt{\frac{\hbar \omega Z_0}{2}} \left\{ \hat{a}_{\text{out}}[\omega] e^{-i\omega(t+\frac{z}{v})} - (\hat{a}_{\text{out}}[\omega])^\dagger e^{i\omega(t+\frac{z}{v})} \right\}. \tag{3.81}$$

The right and left propagating modes are

$$\hat{a}_{\text{in}}[\omega] = 2\pi \sqrt{\frac{v}{2d}} \sum_{k>0} \hat{a}_k \delta(\omega - \omega_k), \tag{3.82}$$

$$\hat{a}_{\text{out}}[\omega] = 2\pi \sqrt{\frac{v}{2d}} \sum_{k<0} \hat{a}_k \delta(\omega - \omega_k), \tag{3.83}$$

and here we rewrite the summation of n into summation of k. By taking the continuum limit $d \to \infty$, the only non-zero commutator of the operators $\hat{a}_{\text{in}}[\omega]$ and $\hat{a}_{\text{out}}[\omega]$ are

$$\left[\hat{a}_{\text{in}}[\omega], (\hat{a}_{\text{in}}[\omega'])^\dagger \right] = 2\pi \delta[\omega - \omega'], \tag{3.84}$$

$$\left[\hat{a}_{\text{out}}[\omega], (\hat{a}_{\text{out}}[\omega'])^\dagger \right] = 2\pi \delta[\omega - \omega']. \tag{3.85}$$

If the transmission line is in thermal equilibrium, then

$$\left\langle (\hat{a}_{\text{in}}[\omega])^\dagger \hat{a}_{\text{in}}[\omega'] \right\rangle = 2\pi \delta[\omega - \omega'] n^{\text{th}}[\omega], \tag{3.86}$$

where $n^{\text{th}}[\omega] = 1/\left(\exp \frac{\hbar \omega}{k_B T} - 1 \right)$ is the Bose-Einstein distribution.

After quantizing the field, we can calculate the quantum statistics of the voltage. The noise spectrum of the input voltage is

$$\bar{S}_{V_{\mathrm{in}}V_{\mathrm{in}}}[\omega] = \int dt \langle \{\hat{V}_{\mathrm{in}}(t), \hat{V}_{\mathrm{in}}(0)\}\rangle e^{i\omega t} = \frac{Z_0}{2}\hbar|\omega|\coth\left(\frac{\hbar|\omega|}{2k_BT}\right). \tag{3.87}$$

For $k_BT \gg \hbar\omega$, Eq. (3.87) reduces to the classical result Eq. (3.23). As discussed in subsection 2.2.5, a semi-infinite transmission line is equivalent to a resistor, and the quantum noise from a resistor is given by the noise voltage at the open terminal of the transmission line ($V = 2V_{\mathrm{in}}$), and therefore

$$\bar{S}_{VV}[\omega] = 4\bar{S}_{V_{\mathrm{in}}V_{\mathrm{in}}}[\omega] = 2Z_0\hbar|\omega|\coth\left(\frac{\hbar|\omega|}{2k_BT}\right). \tag{3.88}$$

3.5 Quantum Langevin equations and approximations

After individually quantized a LC circuit and a transmission line. In this section, we will combine both results to describe the quantum dynamics of a LC circuit coupled to a transmission line. We will also discuss the connection between the quantum circuit theory and the input-output theory in quantum optics [,].

3.5.1 Quantum Langevin equation of a LC circuit

The quantum dynamics of a LC circuit couple to a transmission line is described by the quantum Langevin equation, which is given by replacing the magnetic flux Φ and the input voltage V_{in} in the Langevin equation (3.18) with the corresponding operators (3.63) and (3.80), which gives

$$\frac{d^2\hat{\Phi}}{dt^2} + \frac{1}{Z_0C_{\mathrm{tot}}}\frac{d\hat{\Phi}}{dt} + \frac{\hat{\Phi}}{LC_{\mathrm{tot}}} = \frac{1}{C_{\mathrm{tot}}}\hat{I}_n(t), \tag{3.89}$$

where $\hat{I}_n(t) = \frac{2}{Z_0}\hat{V}_{\mathrm{in}}(t)$. For high Q resonator as the case of our experiment, several approximations can be made to simplify Eq. (3.89). In our experiment, we only focus on the dynamics in a narrow bandwidth around the cavity resonance frequency. In this case, we can rewrite the input voltage (3.80) to

$$\hat{V}_{\text{in}}(t) \simeq -i\sqrt{\frac{\hbar Z \omega_c}{2}} \left(e^{-i\omega_c t} \int_{-\omega_c}^{\infty} \frac{d\omega}{2\pi} \hat{a}_{\text{in}} \left[\omega + \omega_c\right] e^{-i\omega t} - \text{H.c.} \right). \tag{3.90}$$

Since the cavity field only response around the cavity resonance frequency, we can extend the integral from $-\omega_c$ to $-\infty$. Defining the envelope operator of the input field,

$$\hat{a}_{\text{in}}(t) = \int_{-\infty}^{\infty} \frac{d\omega}{2\pi} \hat{a}_{\text{in}} \left[\omega + \omega_c\right] e^{-i\omega t}, \tag{3.91}$$

which satisfies the commutation relation

$$\left[\hat{a}_{\text{in}}(t), \hat{a}_{\text{in}}^{\dagger}(t') \right] = \delta(t - t'), \tag{3.92}$$

and the input voltage is simplified to

$$\hat{V}_{\text{in}}(t) = -i\sqrt{\frac{\hbar Z \omega_c}{2}} \left(\hat{a}_{\text{in}}(t) e^{-i\omega_c t} - \hat{a}_{\text{in}}^{\dagger}(t) e^{i\omega_c t} \right). \tag{3.93}$$

The flux operator (3.63) in a frame rotating at ω_c is given by

$$\hat{\Phi}(t) = \sqrt{\frac{\hbar Z}{2}} \left(\hat{a}(t) e^{-i\omega_c t} + \hat{a}(t)^{\dagger} e^{i\omega_c t} \right), \tag{3.94}$$

where $\hat{a}(t)$ and $\hat{a}(t)^{\dagger}$ represent the envelope of the field. If we substitute Eq. (3.93) and Eq. (3.94) into the quantum Langevin equation (3.89) and separate the parts oscillating at $e^{i\omega t}$ and $e^{-i\omega t}$, we obtain

$$\frac{1}{2\omega_c} i \frac{d^2 \hat{a}}{dt^2} + \frac{d\hat{a}}{dt} + \frac{1}{2Q} i \frac{d\hat{a}}{dt} + \frac{\kappa}{2} \hat{a} = \sqrt{\kappa} \hat{a}_{\text{in}}(t). \tag{3.95}$$

By making the slow varying envelope approximation ($\frac{d^2 \hat{a}}{dt^2} \ll \omega_c \frac{d\hat{a}}{dt}$) and considering high quality resonator ($Q \gg 1$), we can reduce Eq. (3.95) to a first order differential equation.

Transforming back to the lab frame, we obtain

$$\frac{d\hat{a}\left(t\right)}{dt} + \left(i\omega_c + \frac{\kappa}{2}\right)\hat{a}\left(t\right) = \sqrt{\kappa}\hat{a}_{\text{in}}\left(t\right).$$ (3.96)

For multi-ports system, the corresponding input-output relation (3.15) is

$$\hat{a}_{\sigma,\text{out}}\left(t\right) = \hat{a}_{\sigma,\text{in}}\left(t\right) - \sqrt{\kappa_\sigma}\hat{a}\left(t\right),$$ (3.97)

where σ is the index of the port. Eq. (3.96) and Eq. (3.97) can also be obtained from the input-output theory in quantum optics, a brief discussion of the input-output theory is given in appendix A.

3.5.2 Quantum Langevin equation of a mechanical oscillator

Similar to the analogue between the classical Langevin equation of a LC circuit and a mechanical oscillator, the quantum Langevin equation of a mechanical oscillator is given by making the analogues (3.37). The resulting quantum Langevin equation is

$$m\frac{d^2\hat{x}\left(t\right)}{dt^2} + m\Gamma_m\frac{d\hat{x}\left(t\right)}{dt} + m\omega_m^2\hat{x} = \hat{F}_n\left(t\right).$$ (3.98)

The noise spectrum of the Langevin force $\hat{F}_n$ is

$$\bar{S}_{F_nF_n}\left[\omega\right] = 2m\Gamma_m\hbar|\omega|\coth\left(\frac{\hbar\omega}{2k_BT}\right).$$ (3.99)

For high Q mechanical oscillator, the same approximation in the last section can be made. The quantum Langevin equation (3.98) can be simplified to

$$\frac{d\hat{b}\left(t\right)}{dt} + \left(i\omega_m + \frac{\Gamma_m}{2}\right)\hat{b}\left(t\right) = \sqrt{\kappa}\hat{b}_{\text{in}}\left(t\right),$$ (3.100)

where $\hat{b}\left(t\right)$ is the annihilation operator of the mechanical mode.

3.5.3 Quantum Langevin equations of an electromechanical system

Similar to the quantum Langevin equations of the LC circuit and the mechanical oscillator, the quantum Langevin equation of a cavity electromechanical system is given by
replacing the classical variables in Eqs. (3.43) and (3.44) with the corresponding operators,
which gives

$$\frac{d^2\hat{\Phi}}{dt^2} + \frac{1}{Z_0 C_{\text{tot}}(\hat{x})} \frac{d\hat{\Phi}}{dt} + \frac{\hat{\Phi}}{L C_{\text{tot}}(\hat{x})} = \frac{1}{C_{\text{tot}}} \left[\hat{I}_n(t) + I_d(t) \right], \tag{3.101}$$

$$m\frac{d^2\hat{x}}{dt^2} + m\Gamma_m \frac{d\hat{x}}{dt} + m\omega_m^2 \hat{x} = \hat{F}_n(t) + \hat{F}_{\text{rad}}(t). \tag{3.102}$$

Here we include the external drive current $I_d(t)$ in Eq. (3.101). The second term in the
right hand side of Eq. (3.102) is the radiation pressure from the cavity field, which is given
by $\hat{F}_{\text{rad}}(t) = -\frac{1}{2}\frac{\partial C}{\partial x} \left[\hat{V}(t) \right]^2$. By making the same approximations in the previous sections,
Eqs. (3.101) and (3.102) can be simplified to

$$\frac{d\hat{a}(t)}{dt} + \left[i\omega_c(\hat{x}) + \frac{\kappa}{2} \right] \hat{a}(t) = \sqrt{\kappa}\left[\hat{a}_{\text{in}}(t) + \alpha_{\text{in}}(t)\right], \tag{3.103}$$

$$\frac{d\hat{b}(t)}{dt} + \left(i\omega_m + \frac{\Gamma_m}{2} \right) \hat{b}(t) = \sqrt{\kappa}\hat{b}_{\text{in}}(t) + i\frac{x_{\text{zp}}}{\hbar}\hat{F}_{\text{rad}}(t), \tag{3.104}$$

where $\alpha_{\text{in}}(t)$ is the external drive field. For multi-port cavity $\hat{a}_{\text{in}}(t) = \sum_\sigma \sqrt{\frac{\kappa_\sigma}{\kappa}} \hat{a}_{\sigma,\text{in}}(t)$
and $\alpha_{\text{in}}(t) = \sum_\sigma \sqrt{\frac{\kappa_\sigma}{\kappa}} \alpha_{\sigma,\text{in}}(t)$, where σ is the index of the ports. We ignore the position
dependent of κ in Eq.(3.103) because it is very small compare to ω_c. If we define the
electromechanical coupling,

$$g_0 = -\frac{\partial\omega_c}{\partial x} x_{\text{zp}} = \frac{\omega_c}{2} \frac{1}{C_{\text{tot}}} \frac{\partial C}{\partial x} x_{\text{zp}}, \tag{3.105}$$

and expand the $\omega_c(\hat{x})$ to the first order, the quantum Langevin equations of the electromechanical system can be written to

$$\frac{d\hat{a}(t)}{dt} + \left(i\omega_c + \frac{\kappa}{2} \right) \hat{a}(t) - \sqrt{\kappa}\hat{a}_{\text{in}}(t) = \sqrt{\kappa}\alpha_{\text{in}}(t) + ig_0\left[\hat{b}(t) + \hat{b}^\dagger(t) \right], \tag{3.106}$$

$$\frac{d\hat{b}(t)}{dt} + \left(i\omega_m + \frac{\Gamma_m}{2}\right)\hat{b}(t) - \sqrt{\kappa}\hat{b}_{\text{in}}(t) = ig_0\left[\hat{a}^\dagger\hat{a}(t) + \frac{1}{2}\right]. \tag{3.107}$$

In the following chapter, we will used the quantum Langevin equations (3.106) and (3.107) to study the quantum behavior of the cavity electromechanical system. The 1/2 on the right hand side of Eq. (3.107) is the static radiation pressure from the quantum fluctuation of the cavity field, which have no dynamical effect on the system. We will neglect this contribution in the following discussions.

The quantum Langevin equations (3.106) and (3.107) can also be derived from the the input-output theory in quantum optics. The total Hamiltonian of the electromechanical system is

$$\hat{H} = \hat{H}_{\text{cav}} + \hat{H}_{\text{mech}} + \hat{H}_{\text{Int}} + \hat{H}_{\text{dr}}, \tag{3.108}$$

with

$$\hat{H}_{\text{cav}} = \hbar\omega_c\hat{a}^\dagger\hat{a}, \tag{3.109}$$

$$\hat{H}_{\text{mech}} = \hbar\omega_m\hat{b}^\dagger\hat{b}, \tag{3.110}$$

$$\hat{H}_{\text{Int}} = -\hbar g_0\hat{a}^\dagger\hat{a}\left(\hat{b}^\dagger + \hat{b}\right), \tag{3.111}$$

$$\hat{H}_{\text{dr}} = \hbar\sum_{\sigma}\sqrt{\kappa_\sigma}\left[\alpha_{\text{in}}^*(t)\,\hat{a} + \alpha_{\text{in}}(t)\,\hat{a}^\dagger\right], \tag{3.112}$$

where $\hat{H}_{\text{cav}}$ is the Hamiltonian of the optical or microwave cavity. ω_c is the cavity resonance frequency and $\hat{a}\left(\hat{a}^\dagger\right)$ is the annihilation (creation) operator of the cavity photon. $\hat{H}_{\text{mech}}$ is the Hamiltonian of the mechanical oscillator, ω_m is the resonance frequency of the mechanical oscillator, $\hat{b}\left(\hat{b}^\dagger\right)$ is the annihilation (creation) operator of the phonon, $\hat{H}_{\text{Int}}$ is the electromechanical interaction with electromechanical coupling g_0, and $\hat{H}_{\text{dr}}$ is the Hamiltonian of the external drives. Then the quantum Langevin equations (3.106) and (3.107) are given by the Heisenberg equations

$$\dot{\hat{a}} = \frac{i}{\hbar}\left[\hat{H}',\hat{a}\right] - \frac{\kappa}{2}\hat{a} + \sqrt{\kappa}\hat{d}_{\text{in}}, \tag{3.113}$$

$$\dot{\hat{b}} = \frac{i}{\hbar}\left[\hat{H}',\hat{b}\right] - \frac{\Gamma_m}{2}\hat{b} + \sqrt{\Gamma_m}\hat{c}_{\text{in}}. \tag{3.114}$$

Chapter 4

Cavity electromechanics with bichromatic drives

In the previous chapter, we have introduced a quantum theory to describe an electrome-chanical system. Here we will apply this theory to study the dynamics of an electrome-chanical system under bichromatic drives. In this work, we study the stationary behavior of the system through the measurement of the transmission spectrum and the noise spectrum. In this section, we will calculate these spectra with the quantum Langevin equations. The behavior of the system under some special drive configurations will be discussed. The derivations in this chapter represent a combination of results from our theory collaborators (Andreas Kronwald, Anja Metelmann and Aashish Clerk) and my own calculations.

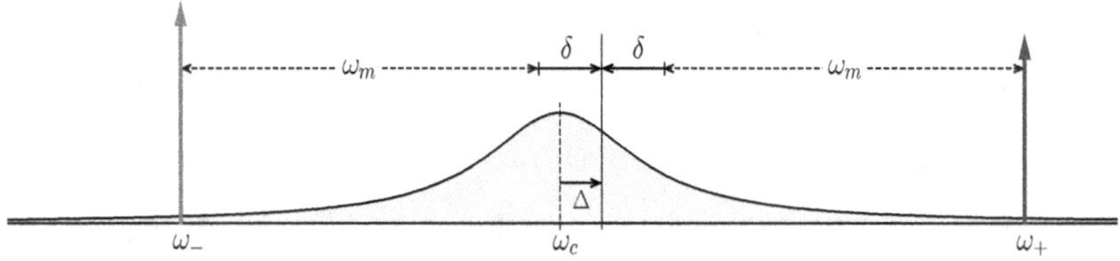

Figure 4.1: Schematic of the pump configuration in the frequency domain. The blue Lorentzian represents the cavity resonance, and the red and blue arrows represent the drive tones.

4.1   Standard formalism and general solution

In this section, we study a two port cavity electromechanical system under bichromatic drive. In the following, we will discuss the standard formalism and approximations to calculate the noise spectra of the system. Consider driving from the input port (L) of the cavity, the Hamiltonian of the drive is

$$\hat{H}_{\mathrm{dr}} = \hbar\sqrt{\kappa_L}\sum_{\nu=\pm}\alpha_\nu\left(\hat{a}e^{i\omega_\nu t} + \hat{a}^\dagger e^{-i\omega_\nu t}\right), \tag{4.1}$$

where κ_L is the coupling of the cavity to the input port, $\omega_\pm = \omega_c + \Delta \pm (\omega_m + \delta)$ and $\alpha_\pm$ are frequencies and amplitudes of the red and blue detuned drives. Δ and δ are the detunings of the drives, which are indicated by the schematic Fig. 4.1.

4.1.1   Rotating frame

To simplify the calculation, we transform into a frame with respect to $\hat{H}_0 = \hbar\left(\omega_c + \Delta\right)\hat{a}^\dagger\hat{a} + \hbar\left(\omega_m + \delta\right)\hat{b}^\dagger\hat{b}$ with the unitary transformation $\hat{U} = e^{i\hat{H}_0 t/\hbar}$. The Hamiltonian in the rotating frame is given by

$$\begin{aligned}
\hat{H}' &= \hat{U}\hat{H}\hat{U}^\dagger - \hat{H}_0\\
&= -\hbar\Delta\hat{a}^\dagger\hat{a} - \hbar\delta\hat{b}^\dagger\hat{b} + \hbar g_0\hat{a}^\dagger\hat{a}\left[e^{-i(\omega_m+\delta)t}\hat{b} + e^{i(\omega_m+\delta)t}\hat{b}^\dagger\right]\\
&\quad + \hbar\sqrt{\kappa_L}\sum_{\nu=\pm}\alpha_\nu\left[\hat{a}e^{\nu i(\omega_m+\delta)t} + \hat{a}^\dagger e^{-\nu i(\omega_m+\delta)t}\right].
\end{aligned} \tag{4.2}$$

The quantum Langevin equations in the rotating frame are

$$\begin{aligned}
\dot{\hat{a}} &= \left(i\Delta - \frac{\kappa}{2}\right)\hat{a} + ig_0\hat{a}\left[e^{-i(\omega_m+\delta)t}\hat{b} + e^{i(\omega_m+\delta)t}\hat{b}^\dagger\right]\\
&\quad - i\left[\alpha_+ e^{-i(\omega_m+\delta)t} + \alpha_- e^{i(\omega_m+\delta)t}\right] + \sqrt{\kappa}\hat{a}_{\mathrm{in}},
\end{aligned} \tag{4.3}$$

$$\dot{\hat{b}} = \left(i\delta - \frac{\Gamma_m}{2}\right)\hat{b} + ig_0\hat{a}^\dagger\hat{a}e^{i(\omega_m+\delta)t} + \sqrt{\Gamma_m}\hat{b}_{\mathrm{in}}, \tag{4.4}$$

where $\hat{a}_{\text{in}} = \sum_{\sigma=L,R,I} \sqrt{\frac{\kappa_\sigma}{\kappa}} \hat{a}_{\sigma,\text{in}}$ is the total input noise of the photon bath, where $\hat{a}_{\sigma,\text{in}}$ describes the input fluctuations to the cavity from channel σ with dampling rate κ_σ. $\sigma = L$ and R correspond to the input and the output ports of the cavity, while $\sigma = I$ corresponds to the internal loss. The noise operator $\hat{b}_{\text{in}}$ describes the noise from the phonon bath with intrinsic damping rate Γ_m. The non-zero commutators of the input field operators are

$$\left[\hat{a}_{\sigma,\text{in}}(t), \hat{a}^{\dagger}_{\sigma',\text{in}}(t') \right] = \delta_{\sigma,\sigma'} \delta(t - t'), \tag{4.5}$$

$$\left[\hat{b}_{\text{in}}(t), \hat{b}^{\dagger}_{\text{in}}(t') \right] = \delta(t - t'). \tag{4.6}$$

They have the following statistics

$$\left\langle \hat{a}^{\dagger}_{\sigma,\text{in}}(t) \, \hat{a}_{\sigma',\text{in}}(t') \right\rangle = n^{\text{th}}_\sigma \delta_{\sigma,\sigma'} \delta(t - t'), \tag{4.7}$$

$$\left\langle \hat{b}^{\dagger}_{\text{in}}(t') \, \hat{b}_{\text{in}}(t) \right\rangle = n^{\text{th}}_m \delta(t - t'), \tag{4.8}$$

where n^{th}_σ is the photon occupation at port σ, n^{th}_m is the phonon bath occupation.

4.1.2 Standard linearization

For the electromechanical system considered in this work, the single phonon electromechanical coupling is very weak ($g_0 \ll \kappa$). In order to enhance the electromechanical interaction, we drive the cavity with strong classical coherent fields. To simplify the calculation, we apply a shift to the cavity field and phonon field

$$\hat{a} \rightarrow \bar{\alpha} + \hat{a}, \tag{4.9}$$

$$\hat{b} \rightarrow \bar{\beta} + \hat{b}, \tag{4.10}$$

where $\bar{\alpha}$ and $\bar{\beta}$ are the classical solutions of the quantum Langevin equations (4.3) and (4.4) that are first order in g_0, i.e.,

$$\dot{\bar{\alpha}} + \left(\frac{\kappa}{2} - i\Delta \right) \bar{\alpha} = -i\sqrt{\kappa_L} \left[\alpha_+ e^{-i(\omega_m+\delta)t} + \alpha_- e^{i(\omega_m+\delta)t} \right], \tag{4.11}$$

$$\dot{\bar{\beta}} + \left(\frac{\Gamma_m}{2} - i\delta\right)\bar{\beta} = ig_0\,|\bar{\alpha}|^2\,e^{i(\omega_m+\delta)t}. \tag{4.12}$$

The steady state solutions are given by

$$\bar{\alpha} = \bar{\alpha}_+ e^{-i(\omega_m+\delta)t} + \bar{\alpha}_- e^{i(\omega_m+\delta)t}, \tag{4.13}$$

$$\bar{\beta} = -i\frac{g_0\,|\bar{\alpha}|^2}{\left(\frac{\Gamma_m}{2} - i\delta\right)}e^{i(\omega_m+\delta)t}, \tag{4.14}$$

with

$$\bar{\alpha}_\pm = -i\frac{\sqrt{\kappa_L}}{\frac{\kappa}{2} - i\left(\Delta \pm \omega_m \pm \delta\right)}\alpha_\pm. \tag{4.15}$$

The intracavity photon numbers corresponding to the two drives are given by $n_p^\pm = |\bar{\alpha}_\pm|^2$.

Linearizing the Langevin equations into first order in the small fluctuations $\hat{a}$ and $\hat{b}$, and subtracting the classical parts, we obtain the dynamics of the quantum fluctuations

$$\dot{\hat{a}} + \left(\frac{\kappa}{2} - i\Delta\right)\hat{a} - \sqrt{\kappa}\hat{a}_{\mathrm{in}} = i\left(G_-\hat{b} + G_+\hat{b}^\dagger\right) +$$
$$i\left[G_+\hat{b}e^{-2i(\Omega+\delta)t} + G_-\hat{b}^\dagger e^{2i(\Omega+\delta)t}\right], \tag{4.16}$$

$$\dot{\hat{b}} + \left(\frac{\Gamma_m}{2} - i\delta\right)\hat{b} - \sqrt{\Gamma_m}\hat{b}_{\mathrm{in}} = i\left(G_-^*\hat{a} + G_+\hat{a}^\dagger\right) +$$
$$i\left(G_+^* e^{2i(\Omega+\delta)t}\hat{a} + G_- e^{2i(\Omega+\delta)t}\hat{a}^\dagger\right), \tag{4.17}$$

where $G_\pm = g_0\bar{\alpha}_\pm$ are the enhanced electromechanical couplings. Here we neglect the cavity frequency shift $g_0\left[e^{-i(\omega_m+\delta)t}\bar{\beta} + e^{i(\omega_m+\delta)t}\bar{\beta}^*\right]$, which is very small in the parameters regime of our experiment.

4.1.3 Rotating wave appoximation

From the resulting linearized quantum Langevin equations (4.16) and (4.17), we find that the linearized electromechanical system can be described by the effective Hamiltonian

$$\hat{H}_{\mathrm{lin}} = \hat{H}_{\mathrm{RWA}} + \hat{H}_{\mathrm{CR}}, \tag{4.18}$$

where

$$\hat{H}_{\text{RWA}} = -\hbar\Delta\hat{a}^\dagger\hat{a} - \hbar\delta\hat{b}^\dagger\hat{b} - \hbar\left[\left(G_-\hat{b} + G_+\hat{b}^\dagger\right)a^\dagger + \left(G_-^*\hat{b}^\dagger + G_+^*\hat{b}\right)\hat{a}\right] \tag{4.19}$$

describes the resonant part of the linearized electromechanical interaction whereas

$$\hat{H}_{\text{CR}} = -\hbar\left[G_+\hat{b}e^{-2i(\omega_m+\delta)t} + G_-\hat{b}^\dagger e^{2i(\omega_m+\delta)t}\right]\hat{a}^\dagger$$
$$-\hbar\left[G_+^*\hat{b}^\dagger e^{2i(\omega_m+\delta)t} + G_-^*\hat{b}e^{-2i(\omega_m+\delta)t}\right]\hat{a}$$

describes the off-resonant electromechanical interactions. The linearized quantum Langevin equations (4.16) and (4.17) can be obtained by

$$\dot{\hat{a}} = \frac{i}{\hbar}\left[\hat{H}_{\text{lin}}, \hat{a}\right] - \frac{\kappa}{2}\hat{a} + \sqrt{\kappa}\hat{a}_{\text{in}}, \tag{4.20}$$

$$\dot{\hat{b}} = \frac{i}{\hbar}\left[\hat{H}_{\text{lin}}, \hat{b}\right] - \frac{\Gamma_m}{2}\hat{b} + \sqrt{\Gamma_m}\hat{b}_{\text{in}}. \tag{4.21}$$

In good cavity limit ($\omega_m \gg \kappa$), the off-resonant part rotate much faster than the time scale of the cavity field. Therefore the average effect of the non-resonance terms are very small. In this case, we can neglect the off-resonance term $\hat{H}_{\text{CR}}$ in this limit, which is the rotating wave approximation (RWA). Then the linearized quantum Langevin equations are simplified to

$$\dot{\hat{a}} + \left(\frac{\kappa}{2} - i\Delta\right)\hat{a} = \sqrt{\kappa}\hat{a}_{\text{in}} + i\left(G_-\hat{b} + G_+\hat{b}^\dagger\right), \tag{4.22}$$

$$\dot{\hat{b}} + \left(\frac{\Gamma_m}{2} - i\delta\right)\hat{b} = \sqrt{\Gamma_m}\hat{b}_{\text{in}} + i\left(G_-^*\hat{a} + G_+\hat{a}^\dagger\right). \tag{4.23}$$

Corrections due to the off-resonance is discussed in appendix B.

4.1.4 Output spectra

In the experiment, we study the dynamics of the electromechanical system through the transmission spectrum and the noise spectrum of the cavity. To calculate these spectra, we transform the quantum Langevin equations into frequency domain with the Fourier transformation

$$\hat{A}[\omega] = \int dt \hat{A}(t) e^{i\omega t}. \tag{4.24}$$

$$\hat{A}^\dagger[\omega] = \int dt \hat{A}^\dagger(t) e^{i\omega t} = \left(\hat{A}[-\omega]\right)^\dagger. \tag{4.25}$$

The transformed quantum Langevin equations are

$$
\begin{aligned}
\chi_c^{-1}[\omega + \Delta]\,\hat{a}[\omega] &= \sqrt{\kappa}\,\hat{a}_{\text{in}}[\omega] + i\left(G_-\hat{b}[\omega] + G_+\hat{b}^\dagger[\omega]\right), \\
\chi_c^{-1}[\omega - \Delta]\,\hat{a}^\dagger[\omega] &= \sqrt{\kappa}\,\hat{a}_{\text{in}}^\dagger[\omega] - i\left(G_-^*\hat{b}^\dagger[\omega] + G_+^*\hat{b}[\omega]\right), \\
\chi_m^{-1}[\omega + \delta]\,\hat{b}[\omega] &= \sqrt{\Gamma_m}\,\hat{b}_{\text{in}}[\omega] + i\left(G_-^*\hat{a}[\omega] + G_+\hat{a}^\dagger[\omega]\right), \\
\chi_m^{-1}[\omega - \delta]\,\hat{b}^\dagger[\omega] &= \sqrt{\Gamma_m}\,\hat{b}_{\text{in}}^\dagger[\omega] - i\left(G_-\hat{a}^\dagger[\omega] + G_+^*\hat{a}[\omega]\right),
\end{aligned}
\tag{4.26}
$$

where $\chi_c[\omega] = \frac{1}{\kappa/2 - i\omega}$ and $\chi_m[\omega] = \frac{1}{\Gamma_m/2 - i\omega}$ are the susceptibilities of the cavity field and the mechanical motion.

If we define the vectors $\hat{D} = \left(\hat{a}, \hat{a}^\dagger, \hat{b}, \hat{b}^\dagger\right)^T$, $\hat{D}_{\text{in}} = \left(\hat{a}_{\text{in}}, \hat{a}_{\text{in}}^\dagger, \hat{b}_{\text{in}}, \hat{b}_{\text{in}}^\dagger\right)^T$ and the diagonal matrix $L = \text{diag}\left(\sqrt{\kappa}, \sqrt{\kappa}, \sqrt{\Gamma_m}, \sqrt{\Gamma_m}\right)$. We can write the quantum Langevin equations into matrix form

$$\left(\chi[\omega]\right)^{-1}\hat{D}[\omega] = L \cdot \hat{D}_{\text{in}}[\omega], \tag{4.27}$$

with the susceptibility matrix

$$
\left(\chi[\omega]\right)^{-1} \equiv
\begin{pmatrix}
\chi_c^{-1}[\omega + \Delta] & 0 & -iG_- & -iG_+ \\
0 & \chi_c^{-1}[\omega - \Delta] & iG_+^* & iG_-^* \\
-iG_-^* & -iG_+ & \chi_m^{-1}[\omega + \delta] & 0 \\
iG_+^* & iG_- & 0 & \chi_m^{-1}[\omega - \delta]
\end{pmatrix},
\tag{4.28}
$$

and the solution of the cavity fields and the phonon fields are given by

$$\hat{D}[\omega] = \chi[\omega] \cdot L \cdot \hat{D}_{\text{in}}[\omega]. \tag{4.29}$$

In the experiment, we measure the transmission signal through the output port (R) of the cavity, which is given by the input-output relation

$$\hat{a}_{R,\text{out}}[\omega] = \hat{a}_{R,\text{in}}[\omega] - \sqrt{\kappa_R}\,\hat{a}[\omega]. \tag{4.30}$$

Substituting the solution of the intracavity field $\hat{a}[\omega]$ from Eq. (4.29) into Eq. (4.30), we obtain the solution of the output field

$$\hat{d}_{R,\text{out}}[\omega] = \hat{d}_{R,\text{in}}[\omega] - \sqrt{\kappa_R\kappa}\,(\chi[\omega])_{11}\,\hat{d}_{\text{in}}[\omega] - \sqrt{\kappa_R\kappa}\,(\chi[\omega])_{12}\,\hat{d}_{\text{in}}^{\dagger}[\omega]$$
$$- \sqrt{\kappa_R\Gamma_m}\,(\chi[\omega])_{13}\,\hat{c}_{\text{in}}[\omega] - \sqrt{\kappa_R\Gamma_m}\,(\chi[\omega])_{14}\,\hat{c}_{\text{in}}^{\dagger}[\omega]. \tag{4.31}$$

Note that here we write the field in the rotating frame, and to transform it back to the lab frame, one simply need to shift the origin to $\omega_c + \Delta$, i.e., $\omega \to \omega - \omega_c - \Delta$.

4.1.4.1 Complex transmission spectrum

One of the most important spectra in the experiment is the complex transmission spectrum (S_{21}), from which we can precisely extract the resonance frequencies (ω_c, ω_m), the linewidths (κ, Γ_m) and the enhanced electromechanical couplings ($G_\pm$) of the electromechanical system. In the experiment, a vector network analyzer (VNA) is used to send a weak probe tone $\alpha_{L,\text{in}}$ from the input port (L) of the device and compare it with the corresponding transmitted signal $\alpha_{R,\text{out}}$ from the output port (R). The VNA sweeps the frequency of the probe to measure the complex transmission S_{21} as a function of frequency. The transmission spectrum can be theoretically calculated by replacing the operators with the corresponding coherent classical fields (i.e, $\hat{a}_{L,\text{in}}[\omega] \to \alpha_{L,\text{in}}[\omega]$ and the other equal to zero), which gives

$$S_{21}[\omega] = \frac{\alpha_{R,\text{out}}}{\alpha_{L,\text{in}}} = -\sqrt{\kappa_L\kappa_R}\,(\chi[\omega])_{11}. \tag{4.32}$$

4.1.4.2 Output noise spectrum

Another important spectrum is the noise spectrum of the microwave field emitted from the output port (R) of the device. From the noise spectrum, we can extract the cavity occupation n_c^{th} and the mechanical occupation n_m^{th} from the noise spectrum. To measure the output noise spectrum, we perform a linear detection with a spectrum analyzer, which is essentially equivalent to a diode plus filter [ ]. The diode serves as a large bandwidth square-law power detector that is sensitive to the integrated voltage noise over the full frequency domain. To selectively measure the noise power at a certain frequency ω_0, a well-behaved and normalized bandpass filter function $f[\omega]$, which is peaked at $\pm\omega_0$, is used to pick up the power at the designated frequency. The voltage at the output of the filter and the input of the diode is

$$V_f[\omega] = f[\omega] V_{\text{out}}[\omega], \tag{4.33}$$

where $V_{\text{out}}[\omega]$ is the output field from the device. With the filter in place, the diode output is proportional to

$$I = \int_{-\infty}^{\infty} d\omega |f[\omega]|^2 S_{V_{\text{out}} V_{\text{out}}}[\omega]. \tag{4.34}$$

Since the filter function is real value in time domain, its Fourier transform is a symmetric in the frequency domain. For narrow band filter, the filter function can approximated by

$$|f[\omega]|^2 = \frac{1}{2} [\delta(\omega - \omega_0) + \delta(\omega + \omega_0)]. \tag{4.35}$$

Therefore, the measured noise power at the diode is proportional to the symmetric noise spectrum

$$I = \frac{1}{2} \left(S_{V_{\text{out}} V_{\text{out}}}[\omega_0] + S_{V_{\text{out}} V_{\text{out}}}[-\omega_0] \right) = \bar{S}_{V_{\text{out}} V_{\text{out}}}[\omega_0]. \tag{4.36}$$

With the solution (4.31), the noise spectrum from the output port of the cavity can be calculated by

$$\bar{S}_{II}[\omega] = \frac{1}{2} \int dt \left\langle \left\{ \hat{I}(0), \hat{I}(t) \right\} \right\rangle e^{i\omega t} = \frac{1}{2} \int \frac{d\omega'}{2\pi} \left\langle \left\{ \hat{I}[\omega'], \hat{I}[\omega] \right\} \right\rangle, \tag{4.37}$$

and here we express the spectrum in terms of quanta, i.e.,

$$\hat{I} = i\sqrt{\frac{2}{\hbar Z \omega_c}} \hat{V}_{\text{out}} = -i \left(\hat{a}_{R,\text{out}} - \hat{a}_{R,\text{out}}^\dagger \right). \tag{4.38}$$

Since we can only access the positive frequency of the spectrum, the spectrum can be simplified to

$$\bar{S}_{II}[\omega] = \frac{1}{2} \int \frac{d\omega'}{2\pi} \left\langle \left\{ \hat{a}^{\dagger}_{R,\text{out}}[\omega'], \hat{a}_{R,\text{out}}[\omega] \right\} \right\rangle = \frac{1}{2} + n_R^{\text{th}} + \kappa_R \bar{S}[\omega], \qquad (4.39)$$

with

$$\bar{S}[\omega] = \kappa \left[|(\chi[\omega])_{11}|^2 + |(\chi[\omega])_{12}|^2 \right] \left(n_c^{\text{th}} + \frac{1}{2} \right)$$

$$+ \Gamma_m \left[|(\chi[\omega])_{13}|^2 + |(\chi[\omega])_{14}|^2 \right] \left(n_m^{\text{th}} + \frac{1}{2} \right) - 2\text{Re} \left\{ (\chi[\omega])_{11} \left(n_R^{\text{th}} + \frac{1}{2} \right) \right\}.$$

4.1.5 Quadrature spectrums of the electromechanical system

In addition to the output field of the cavity, another useful physical quantities are the quadratures of the cavity field and the mechanical motion, which are particularly useful when considering squeezed state. The quadratures of the cavity field and the mechanical motion are defined by

$$\hat{U}_1 = \hat{a}^{\dagger} + \hat{a}, \quad \hat{U}_2 = i\left(\hat{a}^{\dagger} - \hat{a}\right),$$
$$\hat{X}_1 = \hat{b}^{\dagger} + \hat{b}, \quad \hat{X}_2 = i\left(\hat{b}^{\dagger} - \hat{b}\right). \qquad (4.40)$$

The corresponding input operators satisfy the following relations:

$$
\begin{aligned}
\left\langle \hat{X}_{1,\text{in}}(t)\,\hat{X}_{1,\text{in}}(t') \right\rangle &= \left(2n_m^T + 1\right)\delta(t - t'), & \left\langle \hat{X}_{1,\text{in}}(t)\,\hat{X}_{2,\text{in}}(t') \right\rangle &= i\delta(t - t'), \\
\left\langle \hat{X}_{2,\text{in}}(t)\,\hat{X}_{2,\text{in}}(t') \right\rangle &= \left(2n_m^T + 1\right)\delta(t - t'), & \left\langle \hat{X}_{2,\text{in}}(t)\,\hat{X}_{1,\text{in}}(t') \right\rangle &= -i\delta(t - t'), \\
\left\langle \hat{U}_{1,\text{in}}(t)\,\hat{U}_{1,\text{in}}(t') \right\rangle &= \left(2n_c^T + 1\right)\delta(t - t'), & \left\langle \hat{U}_{1,\text{in}}(t)\,\hat{U}_{2,\text{in}}(t') \right\rangle &= i\delta(t - t'), \\
\left\langle \hat{U}_{2,\text{in}}(t)\,\hat{U}_{2,\text{in}}(t') \right\rangle &= \left(2n_c^T + 1\right)\delta(t - t'), & \left\langle \hat{U}_{2,\text{in}}(t)\,\hat{U}_{1,\text{in}}(t') \right\rangle &= -i\delta(t - t').
\end{aligned}
$$
$$(4.41)$$

The quantum Langevin equations of the quadratures are

$$\dot{\hat{U}}_1 + \frac{\kappa}{2}\hat{U}_1 = \sqrt{\kappa}\hat{U}_{1,\text{in}} - \Delta\hat{U}_2 - \bar{G}\hat{X}_2,$$

$$\dot{\hat{U}}_2 + \frac{\kappa}{2}\hat{U}_2 = \sqrt{\kappa}\hat{U}_{2,\text{in}} + \Delta\hat{U}_1 + \tilde{G}\hat{X}_1,$$

$$\dot{\hat{X}}_1 + \frac{\Gamma_m}{2}\hat{X}_1 = \sqrt{\Gamma_m}\hat{X}_{1,\text{in}} - \delta\hat{X}_2 - \bar{G}\hat{U}_2, \qquad (4.42)$$

$$\dot{\hat{X}}_2 + \frac{\Gamma_m}{2}\hat{X}_2 = \sqrt{\Gamma_m}\hat{X}_{2,\text{in}} + \delta\hat{X}_1 + \tilde{G}\hat{U}_1,$$

where $\bar{G} = G_- - G_+$ and $\tilde{G} = G_- + G_+$. In frequency domain, the quantum Langevin equations are

$$\chi_c^{-1}[\omega]\,\hat{U}_1[\omega] = \sqrt{\kappa}\hat{U}_{1,\text{in}}[\omega] - \Delta\hat{U}_2[\omega] - \bar{G}\hat{X}_2[\omega],$$

$$\chi_c^{-1}[\omega]\,\hat{U}_2[\omega] = \sqrt{\kappa}\hat{U}_{2,\text{in}}[\omega] + \Delta\hat{U}_1[\omega] + \tilde{G}\hat{X}_1[\omega],$$

$$\chi_m^{-1}[\omega]\,\hat{X}_1[\omega] = \sqrt{\Gamma_m}\hat{X}_{1,\text{in}}[\omega] - \delta\hat{X}_2[\omega] - \bar{G}\hat{U}_2[\omega],$$

$$\chi_m^{-1}[\omega]\,\hat{X}_2[\omega] = \sqrt{\Gamma_m}\hat{X}_{2,\text{in}}[\omega] + \delta\hat{X}_1[\omega] + \tilde{G}\hat{U}_1[\omega].$$

$$(4.43)$$

If we define the vectors $\hat{Q} = \left(\hat{U}_1, \hat{U}_2, \hat{X}_1, \hat{X}_2\right)^T$, $\hat{Q}_{\text{in}} = \left(\hat{U}_{1,\text{in}}, \hat{U}_{2,\text{in}}, \hat{X}_{1,\text{in}}, \hat{X}_{2,\text{in}}\right)^T$, then we can write the quantum Langevin equations in matrix form

$$\left(\chi_Q[\omega]\right)^{-1}\hat{Q}[\omega] = L \cdot \hat{Q}_{\text{in}}[\omega],$$

$$(4.44)$$

with the susceptibility matrix

$$\left(\chi_Q[\omega]\right)^{-1} \equiv \begin{pmatrix} \chi_c^{-1}[\omega] & \Delta & 0 & \bar{G} \\ -\Delta & \chi_c^{-1}[\omega] & -\tilde{G} & 0 \\ 0 & \bar{G} & \chi_m^{-1}[\omega] & \delta \\ -\tilde{G} & 0 & -\delta & \chi_m^{-1}[\omega] \end{pmatrix}.$$

$$(4.45)$$

The solution of the cavity quadratures and mechanical quadratures are given by

$$\hat{Q}[\omega] = \chi_Q[\omega] \cdot L \cdot \hat{Q}_{\text{in}}[\omega].$$

$$(4.46)$$

The quadrature noise spectra of the cavity and the mechanics can be calculated with the solutions (4.46), which are given by

$$\begin{aligned}
\bar{S}_{Q_i}[\omega] &= \frac{1}{2}\int \frac{d\omega'}{2\pi}\left\langle\left\{\hat{Q}_i[\omega'], \hat{Q}_i[\omega]\right\}\right\rangle \\
&= \kappa\left\{\left|\left(\chi_Q[\omega]\right)_{i1}\right|^2 + \left|\left(\chi_Q[\omega]\right)_{i2}\right|^2\right\}\left(2n_c^{\text{th}} + 1\right) \\
&\quad + \Gamma_m\left\{\left|\left(\chi_Q[\omega]\right)_{i3}\right|^2 + \left|\left(\chi_Q[\omega]\right)_{i4}\right|^2\right\}\left(2n_m^{\text{th}} + 1\right).
\end{aligned}$$

$$(4.47)$$

4.2 Single tone electromechanics

Having introduced a generic quantum theory to describe an electromechanical system under bichromatic drives. In the following sections, we will consider some special drive configurations and discuss the effects of the radiation backaction. In this section, we will consider an electromechanical system driven by a single red-detuned tone and a single blue-detuned tone.

4.2.1 Single red-detuned tone

We first consider the behavior of an electromechanical system driven by a single red-detuned tone at $\omega_- = \omega_c - \omega_m$. In this case, the Hamiltonian is given by setting $G_+ = 0$ and $\Delta = \delta = 0$ in Eq. (4.19), which yields

$$\hat{H} = -\hbar G_- \left(\hat{b}\hat{a}^\dagger + \hat{b}^\dagger \hat{a} \right).$$ (4.48)

The quantum Langevin equations in the frequency space are

$$\chi_c^{-1}[\omega]\,\hat{a}\,[\omega] = \sqrt{\kappa}\hat{a}_{\text{in}}[\omega] + iG_-\hat{b}[\omega],$$ (4.49)

$$\chi_m^{-1}[\omega]\,\hat{b}\,[\omega] = \sqrt{\Gamma_m}\hat{b}_{\text{in}}[\omega] + iG_-\hat{a}[\omega].$$ (4.50)

Mechanical noise spectrum: sideband cooling

To examine the effects of the dynamical backaction to the mechanical motion, we calculate the spectral density of the mechanical motion. Substituting Eq. (4.49) into Eq. (4.50), we can express the phonon annihilation operator in terms of the input fields

$$\hat{b}[\omega] = \sqrt{\Gamma_m}\bar{\chi}_m[\omega]\,\hat{b}_{\text{in}}[\omega] + iG_-\sqrt{\kappa}\chi_c[\omega]\,\bar{\chi}_m[\omega]\,\hat{a}_{\text{in}}[\omega],$$ (4.51)

where

$$\bar{\chi}_m[\omega] = \frac{1}{\Gamma_m/2 - i\omega + G_-^2\chi_c[\omega]}$$ (4.52)

is the effective mechanical susceptibility. The dynamical backaction generated by the driven cavity modifies the dynamics of the mechanical resonator. In weak coupling regime ($\kappa \gg G_-$), the effective mechanical susceptibility can be simplified to

$$\bar{\chi}_m[\omega] = \frac{1}{\left(\Gamma_m + \Gamma_{\rm opt}\right)/2 - i\omega}. \tag{4.53}$$

The dynamical backaction generated by the red-detuned tone damp the mechanical motion with the optical damping rate $\Gamma_{\rm opt} = 4G_-^2/\kappa$.

Using the solution (4.51), we can calculate the mechanical spectrum

$$S_{bb}[\omega] = \int_{-\infty}^{\infty} \frac{d\omega'}{2\pi} \left\langle \hat{b}^\dagger[\omega'] \, \hat{b}[\omega] \right\rangle = 4\frac{\left(\kappa^2 + 4\omega^2\right)\Gamma_m n_m^{\rm th} + 4G_-^2 \kappa n_c^{\rm th}}{\left(4G_-^2 + \Gamma_m \kappa\right)^2 + 4\left(\Gamma_m^2 + \kappa^2 - 8G_-^2\right)\omega^2 + 16\omega^4}. \tag{4.54}$$

The first term is the contribution of the Langevin force from the phonon bath, the second term is the random radiation backaction from the cavity photon. The mechanical occupation is given by the integral of the mechanical spectrum

$$\bar{n}_m = \int_{-\infty}^{\infty} \frac{d\omega}{2\pi} S_{bb}[\omega] = \frac{\Gamma_m\left(\Gamma_{\rm tot} + \kappa\right)}{\Gamma_{\rm tot}\left(\kappa + \Gamma_m\right)} n_m^{\rm th} + \frac{\kappa \Gamma_{\rm opt}}{\Gamma_{\rm tot}\left(\kappa + \Gamma_m\right)} n_c^{\rm th}, \tag{4.55}$$

where $\Gamma_{\rm tot} = \Gamma_m + \Gamma_{\rm opt}$ is the total mechanical linewidth.

In weak coupling regime ($\kappa \gg G_-$), the mechanical spectrum can be simplified to

$$S_{bb}[\omega] = \frac{\Gamma_{\rm tot}}{\left(\Gamma_{\rm tot}/2\right)^2 + \omega^2} \bar{n}_m, \tag{4.56}$$

and the mechanical occupation can be expressed by the detailed balance equation

$$\bar{n}_m = \frac{\Gamma_m}{\Gamma_m + \Gamma_{\rm opt}} n_m^{\rm th} + \frac{\Gamma_{\rm opt}}{\Gamma_m + \Gamma_{\rm opt}} n_c^{\rm th}. \tag{4.57}$$

The driven cavity serves as a cold reservoir to the mechanical oscillator, the dynamical backaction damps the mechanical motion and cools the mechanical oscillator [,   ]. When the optical damping rate dominates ($\Gamma_{\rm opt} \gg \Gamma_m$), the mechanical occupation asymptote to the cavity occupation $n_c^{\rm th}$, which is usually much smaller than the occupation of the phonon bath n_m^{th}. The effect of the counter rotating term to the mechanical occupation can be found in [,   ], which gives

$$\bar{n}_m = \frac{\Gamma_m}{\Gamma_{\text{tot}}} n_m^{\text{th}} + \frac{\Gamma_{\text{opt}}}{\Gamma_{\text{tot}}} \left[n_c^{\text{th}} \left(1 + 2 \left(\frac{\kappa}{4\omega_m} \right)^2 \right) + \left(\frac{\kappa}{4\omega_m} \right)^2 \right]. \tag{4.58}$$

In our experiment, the cavity heats up as we increase the pump power, which produces nonzero n_c^{th}. In our experiment, the sideband factor $\left(\frac{\kappa}{4\omega_m} \right)^2 \simeq 2 \times 10^{-4}$ is very small compared to n_c^{th}. Therefore, the corrections from the counter rotating terms are negligible.

Transmission spectrum: electromechanical induced transparency

The transmission spectrum of an electromechanical system under a red-detuned drive is given by setting $G_+ = 0$ and $\Delta = \delta = 0$ in Eq. (4.32), which gives

$$S_{21}[\omega] = -\frac{\sqrt{\kappa_L \kappa_R}}{\kappa/2 - i\omega + \frac{G^2}{\Gamma_m/2 - i\omega}}. \tag{4.59}$$

In weak coupling regime ($G \ll \kappa$), the effective mechanical linewidth Γ_{tot} is much smaller than the cavity linewidth κ. If we focus on the region around the cavity resonance, the transmission spectrum can be simplified to

$$S_{21}[\omega] = -2\frac{\sqrt{\kappa_R \kappa_L}}{\kappa} \frac{\Gamma_m/2 - i\omega}{\Gamma_{\text{tot}}/2 - i\omega} = -2\frac{\sqrt{\kappa_R \kappa_L}}{\kappa} \left(1 - \frac{\Gamma_{\text{opt}}/2}{\Gamma_{\text{tot}}/2 - i\omega} \right). \tag{4.60}$$

The destructive interference between the mechanical sideband and the probe field generates a Lorentzian dip at $\omega = 0$ in the cavity transmission spectrum. This effect is known as the optomechanical induced transparency (OMIT) [,   ].

Output noise spectrum: noise squashing

The output noise spectrum is given by setting $G_+ = 0$ and $\Delta = \delta = 0$ in Eq. (4.39), which yields

$$\bar{S}_{II}[\omega] = n_R^{\text{th}} + \frac{1}{2} + 4\kappa_R \frac{4G_-^2\Gamma_m\left(n_m^{\text{th}} - n_R^{\text{th}}\right) + \left(\Gamma_m^2 + 4\omega^2\right)\kappa\left(n_c^{\text{th}} - n_R^{\text{th}}\right)}{\left(4G_-^2 + \Gamma_m\kappa\right)^2 + 4\left(\Gamma_m^2 + \kappa^2 - 8G_-^2\right)\omega^2 + 16\omega^4}. \tag{4.61}$$

In weak coupling regime, it can be simplified to

$$\bar{S}_{II}[\omega] = \bar{S}_0[\omega] + \frac{\kappa_R}{\kappa}\Gamma_{\text{opt}} S'_{bb}[\omega]. \tag{4.62}$$

The output noise spectrum consists of the noise spectrum of the up-converted mechanical sideband $S'_{bb}[\omega]$ sits on top of the noise background $\bar{S}_0[\omega]$, which is given by

$$S'_{bb}[\omega] = \frac{\Gamma_{\text{tot}}}{(\Gamma_{\text{tot}}/2)^2 + \omega^2}\left(\bar{n}_m - n_{\text{eff}}\right), \tag{4.63}$$

$$\bar{S}_0[\omega] = n_R^{\text{th}} + \frac{1}{2} + \frac{\kappa_R\kappa}{(\kappa/2)^2 + \omega^2}\left(n_c^{\text{th}} - n_R^{\text{th}}\right), \tag{4.64}$$

where $n_{\text{eff}} = 2n_c^{\text{th}} - n_R^{\text{th}}$. Note that when there is nonzero cavity occupation, the area of the up-converted mechanical sideband is not simply proportional to $\bar{n}_m$, but $\bar{n}_m - n_{\text{eff}}$. This reduction of noise quanta in the mechanical sideband is known as the noise squashing effect [76, 83, 106], which is due to the destructive interference between the cavity noise and the transduction of the radiation pressure noise.

Inferred mechanical spectrum

The inferred mechanical spectrum is given by normalizing the output spectrum by the gain $\frac{\kappa_R}{\kappa}\Gamma_{\text{opt}}$, which is

$$\frac{\bar{S}_{xx,\text{tot}}[\omega]}{x_{\text{zp}}^2} = \frac{\bar{S}_{II}[\omega]}{\frac{\kappa_R}{\kappa}\Gamma_{\text{opt}}} = \frac{\kappa\bar{S}_0[\omega]}{\kappa_R\Gamma_{\text{opt}}} + \frac{\Gamma_{\text{tot}}}{(\Gamma_{\text{tot}}/2)^2 + \omega^2}\left(\bar{n}_m - n_{\text{eff}}\right). \tag{4.65}$$

For ideal cavity where $n_c^{\text{th}} = n_R^{\text{th}} = n_{\text{eff}} = 0$, the inferred mechanical spectrum becomes

$$\frac{\bar{S}_{xx,\text{tot}}[\omega]}{x_{\text{zp}}^2} = \frac{\kappa}{2\kappa_R\Gamma_{\text{opt}}} + \frac{\Gamma_{\text{tot}}}{(\Gamma_{\text{tot}}/2)^2 + \omega^2}\bar{n}_m. \tag{4.66}$$

$$= \frac{\kappa}{2\kappa_R\Gamma_{\text{opt}}} + \left(\frac{\Gamma_m}{\Gamma_{\text{tot}}}\right)\frac{\Gamma_{\text{tot}}}{(\Gamma_{\text{tot}}/2)^2 + \omega^2}n_m^{\text{th}}. \tag{4.67}$$

The Lorentzian part of the spectrum (second term in Eq. (4.66)) has the expected form for phonon emission of a mechanical oscillator with average phonon number $\bar{n}_m$. Note that there is no backaction in Eq. (4.67). At the first glance, it seems violating the quantum limit on the detector noise. However, in this case, the dynamical backaction dampens the mechanical motion and increases the mechanical linewidth. The total mechanical linewidth $\Gamma_{\text{tot}} = \Gamma_m + \Gamma_{\text{opt}}$ is larger than the intrinsic mechanical linewidth Γ_m. As discussed at the end of section 2.2.2, this detector would have no power gain, and therefore there is no restriction to the added noise of the detector.

4.2.2 Single blue-detuned tone

For an electromechanical system driven by a single blue-detuned drive, the Hamiltonian is given by setting $G_- = 0$ and $\Delta = \delta = 0$ in Eq. (4.19), which yields

$$\hat{H} = -\hbar G_+ \left(\hat{b}\hat{a} + \hat{b}^\dagger\hat{a}^\dagger\right). \tag{4.68}$$

The quantum Langevin equations in the frequency space are

$$\chi_c^{-1}[\omega]\,\hat{a}[\omega] = \sqrt{\kappa}\hat{a}_{\text{in}}[\omega] + iG_+\hat{b}^\dagger[\omega], \tag{4.69}$$

$$\chi_m^{-1}[\omega]\,\hat{b}^\dagger[\omega] = \sqrt{\Gamma_m}\hat{b}_{\text{in}}^\dagger[\omega] - iG_+^*\hat{a}[\omega]. \tag{4.70}$$

Mechanical noise spectrum: amplification

Substituting Eq. (4.69) into Eq. (4.70), we obtain

$$\hat{b}^\dagger[\omega] = \sqrt{\Gamma_m}\bar{\chi}_m[\omega]\,\hat{b}_{\text{in}}^\dagger[\omega] - iG_+\sqrt{\kappa}\chi_c[\omega]\,\bar{\chi}_m[\omega]\,\hat{a}_{\text{in}}[\omega], \tag{4.71}$$

where

$$\bar{\chi}_m [\omega] = \frac{1}{\Gamma_m/2 - i\omega - G_+^2 \chi_c [\omega]} \tag{4.72}$$

is the effective mechanical susceptibility. In weak coupling regime ($\kappa \gg G_+$), it can be simplified to

$$\bar{\chi}_m [\omega] = \frac{1}{\left(\Gamma_m - \Gamma_{\mathrm{opt}}\right)/2 - i\omega}. \tag{4.73}$$

The dynamical backaction generated by the blue-detuned tone amplifies the mechanical motion and narrow the mechanical linewidth with the optical anti-damping rate $\Gamma_{\mathrm{opt}} = 4G_+^2/\kappa$. When Γ_{opt} approaches the intrinsic damping rate Γ_m, the mechanical oscillator starts to self-oscillate and becomes unstable. Therefore we only consider weak coupling regime in this subsection.

Using the solution (4.71), the mechanical spectrum is given by

$$S_{b^\dagger b^\dagger} [\omega] = \int_{-\infty}^{\infty} \frac{d\omega'}{2\pi} \left\langle \hat{b} [\omega'] \hat{b}^\dagger [\omega] \right\rangle = \frac{\Gamma_{\mathrm{tot}}}{(\Gamma_{\mathrm{tot}}/2)^2 + \omega^2} (\bar{n}_m + 1), \tag{4.74}$$

where $\Gamma_{\mathrm{tot}} = \Gamma_m - \Gamma_{\mathrm{opt}}$. The effective mechanical occupation is

$$\bar{n}_m = \frac{\Gamma_m}{\Gamma_m - \Gamma_{\mathrm{opt}}} n_m^{\mathrm{th}} + \frac{\Gamma_{\mathrm{opt}}}{\Gamma_m - \Gamma_{\mathrm{opt}}} \left(n_c^{\mathrm{th}} + 1\right). \tag{4.75}$$

In blue-detuned case, the anti-damping induced by the dynamical backaction narrow the mechanical linewidth, therefore amplifying the mechanical motion and increasing the mechanical occupation.

Transmission spectrum: electromechanical induced amplification

The transmission spectrum of an electromechanical system driven by a blue-detuned tone is given by setting $G_- = 0$ and $\Delta = \delta = 0$ in Eq. (4.32), which gives

$$S_{21} [\omega] = -\frac{\sqrt{\kappa_L \kappa_R}}{\kappa/2 - i\omega - \frac{G_+^2}{\Gamma_m/2 - i\omega}}. \tag{4.76}$$

The transmission spectrum around the cavity resonance can be simplified to

$$S_{21}[\omega] = -2\frac{\sqrt{\kappa_R \kappa_L}}{\kappa}\frac{\Gamma_m/2 - i\omega}{\Gamma_{\text{tot}}/2 - i\omega} = -2\frac{\sqrt{\kappa_R \kappa_L}}{\kappa}\left(1 + \frac{\Gamma_{\text{opt}}/2}{\Gamma_{\text{tot}}/2 - i\omega}\right). \tag{4.77}$$

In this case, the mechanical sideband and the probe field interfere constructively and generate a Lorentzian peak at $\omega = 0$ in the cavity transmission spectrum. This effect is known as the optomechanical induced amplification (OMIA) [ ].

Output noise spectrum: noise anti-squashing

The output noise specteum is given by setting $G_- = 0$ and $\Delta = \delta = 0$ in Eq. (4.39). In weak coupling regime, it can be expressed as

$$\bar{S}_{II}[\omega] = \bar{S}_0[\omega] + \frac{\kappa_R}{\kappa}\Gamma_{\text{opt}}S'_{b^\dagger b^\dagger}[\omega] \tag{4.78}$$

with

$$S'_{b^\dagger b^\dagger}[\omega] = \frac{\Gamma_{\text{tot}}}{(\Gamma_{\text{tot}}/2)^2 + \omega^2}\left(\bar{n}_m + 1 + n_{\text{eff}}\right). \tag{4.79}$$

The output noise spectrum consists of the noise spectrum of the up-converted mechanical sideband $S'_{b^\dagger b^\dagger}[\omega]$ sits on top of the noise background $\bar{S}_0[\omega]$. When there is nonzero cavity occupation, the area of the mechanical sideband is not simply proportional to $\bar{n}_m + 1$, but $\bar{n}_m + 1 + n_{\text{eff}}$. The extra noise quanta in the mechanical sideband is due to the constructive interference between the cavity noise and the transduction of the radiation pressure noise, which is known as the noise anti-squashing effect [,].

Inferred mechanical spectrum

The inferred mechanical spectrum is given by normalizing the output spectrum by the gain $\frac{\kappa_R}{\kappa}\Gamma_{\text{opt}}$, which is

$$\frac{\bar{S}_{xx,\text{tot}}[\omega]}{x_{\text{zp}}^2} = \frac{\bar{S}_{II}[\omega]}{\frac{\kappa_R}{\kappa}\Gamma_{\text{opt}}} = \frac{\kappa\bar{S}_0[\omega]}{\kappa_R\Gamma_{\text{opt}}} + \frac{\Gamma_{\text{tot}}}{(\Gamma_{\text{tot}}/2)^2 + \omega^2}\left(\bar{n}_m + 1 + n_{\text{eff}}\right). \tag{4.80}$$

For ideal cavity where $n_c^{\text{th}} = n_R^{\text{th}} = n_{\text{eff}} = 0$, the inferred mechanical spectrum becomes

$$\frac{\bar{S}_{xx,\text{tot}}[\omega]}{x_{\text{zp}}^2} = \frac{\kappa}{2\kappa_R \Gamma_{\text{opt}}} + \frac{\Gamma_{\text{tot}}}{(\Gamma_{\text{tot}}/2)^2 + \omega^2}(\bar{n}_m + 1). \tag{4.81}$$

$$= \frac{\kappa}{2\kappa_R \Gamma_{\text{opt}}} + \left(\frac{\Gamma_m}{\Gamma_{\text{tot}}}\right) \frac{\Gamma_{\text{tot}}}{(\Gamma_{\text{tot}}/2)^2 + \omega^2}\left(n_m^{\text{th}} + n_{\text{BA}} + 1\right), \tag{4.82}$$

where $n_{\text{BA}} = \Gamma_{\text{opt}}/\Gamma_m$ is the backaction in terms of quanta. The Lorentzian part of the spectrum (second term in Eq. (4.81)) has the expected form for phonon absorption of a mechanical oscillator with average phonon number $\bar{n}_m$.

4.2.3 Motional sideband asymmetry

As shown in Eqs. (4.66) and (4.81), the inferred mechanical spectra is asymmetric with respect to the detuning. In red-detuned case ($\omega_c - \omega_p = \omega_m$), the area of the Lorentzian part of the spectrum is equal to the mechanical occupation, i.e., $I_- = \bar{n}_m$. While in blue-detuned case ($\omega_c - \omega_p = -\omega_m$) $I_+ = \bar{n}_m + 1$. This asymmetry reflects the ability of an quantum harmonic oscillator to emit and absorb energy; for a harmonic oscillator at ground state ($\bar{n}_m = 0$), it has no ability to emit energy ($I_- = 0$), but it can absorb energy ($I_+ = 1$).

In this section, we will give an alternative explanation to the sideband asymmetry by using the linear response analysis discussed in section 2.2.2 [ ]. To start the discussion, let's identify the coupling A and the input operator of the detector $\hat{F}$ in the interaction Hamiltonian (2.37). For an electromechanical system driven by a single tone at ω_p, in a frame rotating at ω_p, the linearized interaction Hamiltonian is

$$\hat{H}_{\text{Int}} = -\frac{\hbar G}{x_{\text{zp}}}\hat{x}\left(\hat{a} + \hat{a}^\dagger\right), \tag{4.83}$$

where $G = g_0\bar{\alpha}$ is the enhanced electromechanical coupling with $\bar{\alpha} = -\frac{\sqrt{\kappa}}{\kappa/2 + i\Delta_p}\alpha_p$, $\Delta_p = \omega_c - \omega_p$ is the detuning of the drive. Here we choose α_p such that $\bar{\alpha}$ is real without loss of generality. Then, one can easily identify $A = \frac{\hbar G}{x_{\text{zp}}}$ and $\hat{F} = \hat{a} + \hat{a}^\dagger$. The uncoupled cavity satisfies the quantum Langevin equation

$$\dot{\hat{a}} + \left(\frac{\kappa}{2} + i\Delta_p\right)\hat{a} = -\sqrt{\kappa}\hat{a}_{\text{in}}. \tag{4.84}$$

Here we consider a two port cavity, the output of the detector is the transmitted voltage at the output port, which is

$$\hat{I} = -i\left(\hat{a}_{R,\text{out}} - \hat{a}_{R,\text{out}}^{\dagger}\right). \tag{4.85}$$

The linear-response susceptibility of the detector is given by Eq. (2.39), which gives

$$\chi_{IF}[\omega] = -\frac{1}{\hbar}\sqrt{\kappa_R}\left(\chi_c[\omega] + \chi_c^*[\omega]\right), \tag{4.86}$$

where

$$\chi_c[\omega] = \frac{1}{\kappa/2 - i\left(\omega - \Delta_p\right)} \tag{4.87}$$

is the susceptibility of the cavity fields. The damping induced by the detector is given by Eq. (2.54), which gives

$$\Gamma_{\text{det}}[\omega] = \frac{\kappa}{\hbar^2}x_{\text{zp}}^2\left(|\chi_c[\omega]|^2 - |\chi_c[-\omega]|^2\right). \tag{4.88}$$

The noise correlators of the detector backaction $\hat{F}$ and imprecision $\hat{I}$ are given by Eqs. (2.42), (2.43), and (2.44). For a two port electromechanical system driven by a single tone, they are

$$\bar{S}_{FF}[\omega] = \kappa\left(|\chi_c[\omega]|^2 + |\chi_c[-\omega]|^2\right)\left(n_c^{\text{th}} + \frac{1}{2}\right), \tag{4.89}$$

$$\bar{S}_{II}[\omega] = \left(n_R^{\text{th}} + \frac{1}{2}\right) + \kappa_R\kappa\left(|\chi_c[\omega]|^2 + |\chi_c[-\omega]|^2\right)\left(n_c^{\text{th}} - n_R^{\text{th}}\right), \tag{4.90}$$

$$\bar{S}_{IF}[\omega] = i\sqrt{\kappa_R}\left(\chi_c^*[\omega] - \chi_c[-\omega]\right)\left(n_R^{\text{th}} + \frac{1}{2}\right)$$

$$- i\kappa\sqrt{\kappa_R}\left(|\chi_c[\omega]|^2 - |\chi_c[-\omega]|^2\right)\left(n_c^{\text{th}} + \frac{1}{2}\right). \tag{4.91}$$

The inferred mechancial spectrum is given by Eq. (2.67) with the effective mechanical spectrum

$$\bar{S}_{xx,\text{eff}}[\omega] = |\chi_{xx}[\omega]|^2 \left(\bar{S}_{F_0 F_0}[\omega] + A^2 \bar{S}_{FF}[\omega] \right) - 2\text{Re}\left\{ \chi^*_{xx}[\omega] \bar{S}_{zF}[\omega], \right\} \tag{4.92}$$

where $\chi_{xx}[\omega]$ is the mechanical susceptibility which is given by Eq. (2.59), and $\bar{S}_{zF}[\omega] = \bar{S}_{IF}[\omega]/\chi_{IF}[\omega]$. For $\Delta_p = \pm\omega_m$, the inferred mechancial spectrum (2.67) reduced to Eqs. (4.66) and (4.81). Here we focus on the origin of the sideband asymmetry; note that the only contribution that is asymmetric at Δ_p is the cross correlator $\bar{S}_{zF}[\omega]$, which is given by

$$\bar{S}_{zF}\left[\Delta_p\right] \equiv \frac{\bar{S}_{IF}[\omega]}{\chi_{IF}[\omega]} = \pm i\hbar \left(2n_c^{\text{th}} - n_R^{\text{th}} + \frac{1}{2} \right), \tag{4.93}$$

where the plus sign (minus sign) corresponds to the drive detuning $\Delta_p = +\omega_m$ ($\Delta_p = -\omega_m$). The cross correlator is purely imaginary and changes sign for $\Delta_p = \pm\omega_m$, which gives rise to the noise squashing and anti-squashing effects. It is clear that the asymmetry between the spectra at $\Delta_p = \pm\omega_m$ is entirely due to the detector backaction-imprecision correlation described by $\bar{S}_{zF}$, which is simply the squashing and anti-squashing of the cavity quantum fluctuation.

In addition, as discussed in section 2.2.2, when the cross correlator $\bar{S}_{zF}$ is purely imaginary, the rhs of the noise constraint (2.46) can be reduced below $\hbar^2/4$. It achieves the minimum value of 0 in the special case when $n_c^{\text{th}} = 0$ and $\bar{S}_{zF} = \pm i\hbar/2$, i.e.,

$$\bar{S}_{II}[\omega] \bar{S}_{FF}[\omega] = \left| \bar{S}_{IF}[\omega] \right|^2. \tag{4.94}$$

The detector backaction and the imprecision are perfectly correlated. In this case, the detector does not need to generate any added noise. However, as discussed in [,], the detector does not provide any power gain. Therefore, it doesn't violate any quantum limit on position detection.

4.3 Two-tone electromechanics

After discussed the effects of a single drive tone in an electromechanical system. In this section, we will study the behavior of an electromechanical system under two drive tones. We consider three special pump configurations: the balanced detuned two-tone configuration to examine the measurement backaction, the balanced two-tone configuration to perform the backaction evading measurement of a single mechanical quadrature, and the two-tone reservoir engineering scheme to squeeze the mechanical motion.

4.3.1 Balanced detuned two-tone: Measurement backaction

As discussed in section 2.2.2, in order to have large power gain, the intrinsic mechanical linewidth has to be much larger than the damping induced by the detector. For an electromechanical system in the sideband resolved regime, it can be realized by driving the electromechancial system at $\omega_{\pm} = \omega_c \pm (\omega_m + \delta)$ with equal amplitude. In this section, we consider $\kappa \gg \delta \gg \Gamma_m$. At this drive configuration, the Hamiltonian is given by setting $\Delta = 0$ and $G_- = G_+ = G$ in Eqs. (4.19), which yields

$$\hat{H} = -\hbar\delta\hat{b}^\dagger\hat{b} - \hbar G\left(\hat{b}^\dagger + \hat{b}\right)\left(\tilde{\hat{d}}^\dagger + \hat{d}\right). \tag{4.95}$$

The quantum Langevin equations in frequency space are given by

$$\chi_c^{-1}[\omega]\,\hat{a}[\omega] = \sqrt{\kappa}\hat{a}_{\text{in}}[\omega] + iG\left(\hat{b}[\omega] + \hat{b}^\dagger[\omega]\right), \tag{4.96}$$

$$\chi_m^{-1}[\omega + \delta]\,\hat{b}[\omega] = \sqrt{\Gamma_m}\hat{b}_{\text{in}}[\omega] + iG\left(\hat{a}[\omega] + \hat{a}^\dagger[\omega]\right), \tag{4.97}$$

and their conjugate equations.

Mechanical noise spectrum

We first examine the noise spectrum of the mechanical motion. Substituting Eq. (4.96) and its conjugate equation into Eq. (4.97), the solution of the mechanical annihilation operator can be written as

$$\hat{b}[\omega] = \chi_m[\omega + \delta]\left(\sqrt{\Gamma_m}\,\hat{b}_{in}[\omega] + iG\sqrt{\kappa}\chi_c[\omega]\,\hat{a}_{in}[\omega] + iG\sqrt{\kappa}\chi_c[\omega]\,\hat{a}_{in}^{\dagger}[\omega]\right). \quad (4.98)$$

Since we are driving the system with both the red and the blue detuned tones with equal power, the dynamical backaction induced by the two drive tones cancel each other. The only effect of the radiation backaction is the random radiation pressure that drives the mechanics into random motion and increases the mechanical occupation. The noise spectrum of the mechanical motion is

$$S_{xx}[\omega] = \int_{-\infty}^{\infty} \frac{d\omega'}{2\pi}\langle \hat{x}[\omega']\,\hat{x}[\omega]\rangle = x_{zp}^2\left(S_{bb}[\omega] + S_{b^{\dagger}b^{\dagger}}[\omega]\right)$$

$$= x_{zp}^2\frac{\Gamma_m}{(\Gamma_m/2)^2 + (\omega + \delta)^2}\bar{n}_m + x_{zp}^2\frac{\Gamma_m}{(\Gamma_m/2)^2 + (\omega - \delta)^2}(\bar{n}_m + \beta) \quad (4.99)$$

with the effective mechanical occupation

$$\bar{n}_m = n_m^{th} + n_m^{BA}, \quad (4.100)$$

where $n_m^{BA} = \frac{\Gamma_{opt}}{\Gamma_m}\left(2n_c^{th} + \alpha\right)$ is the measurement backaction interms of quanta. Here we separately designate the vacuum fluctuation of the cavity photon and the phonon by α and β, which are both equal to one. Note that even the cavity is in ground state $\left(n_c^{th} = 0\right)$, and the measurement backaction is not equal to zero. The radiation pressure generated by the quantum fluctuation of the cavity field produces minimum measurement backaction to the mechanics, which is the quantum radiation backaction.

Transmission spectrum

The transmission spectrum of an electromechanical system is given by setting $G_- = G_+ = G$ and $\Delta = 0$ in Eq. (4.32). For $\kappa \gg \Gamma_{opt}$ and $\kappa \gg \delta \gg \Gamma_m$, the transmission spectrum near the cavity resonance is

$$S_{21} = -2\frac{\sqrt{\kappa_R \kappa_L}}{\kappa}\left[1 - \frac{\Gamma_{opt}/2}{\Gamma_{tot}/2 - i(\omega + \delta)} + \frac{\Gamma_{opt}/2}{\Gamma_{tot}/2 - i(\omega - \delta)}\right]. \quad (4.101)$$

The second term comes from the destructive interference between the probe field and the up-converted motional sideband of the red-detuned drive, which reduce the cavity transmission at $\omega = -\delta$. The third term comes from the constructive interference of the probe field and down-converted motional sideband of the blue-detuned drive, which amplify the cavity transmission at $\omega = \delta$.

Output noise spectrum

The noise spectrum of the output field is given by setting $\Delta = 0$ and $G_- = G_+ = G$ in Eq. (4.39). In weak coupling regime, it can be simplified to

$$\bar{S}_{II}[\omega] = \bar{S}_0[\omega] + \frac{\kappa_R}{\kappa}\Gamma_{\text{opt}}\left(S'_{bb}[\omega + \delta] + S'_{b^\dagger b^\dagger}[\omega - \delta]\right). \tag{4.102}$$

The output noise spectrum consists of the noise background $\bar{S}_0[\omega]$, the noise spectrum of the up-converted mechanical sideband $S'_{bb}[\omega]$, and the noise spectrum of the down-converted mechanical sideband $S'_{b^\dagger b^\dagger}[\omega]$, which are given by

$$\bar{S}_0[\omega] = n_R^{\text{th}} + \frac{\alpha}{2} + \frac{\kappa_R\kappa}{(\kappa/2)^2 + \omega^2}\left(n_c^{\text{th}} - n_R^{\text{th}}\right), \tag{4.103}$$

$$S'_{bb}[\omega] = \frac{\Gamma_m}{(\Gamma_m/2)^2 + \omega^2}\left[\left(\bar{n}_m + \frac{\beta}{2}\right) - \left(n_{\text{eff}} + \frac{\alpha}{2}\right)\right], \tag{4.104}$$

$$S'_{b^\dagger b^\dagger}[\omega] = \frac{\Gamma_m}{(\Gamma_m/2)^2 + \omega^2}\left[\left(\bar{n}_m + \frac{\beta}{2}\right) + \left(n_{\text{eff}} + \frac{\alpha}{2}\right)\right], \tag{4.105}$$

where $\Gamma_{\text{opt}} = \frac{4G^2}{\kappa}$, $n_{\text{eff}} = 2n_c^{\text{th}} - n_R^{\text{th}}$, α and β are the quantum fluctuations of the photon and phonon. Similar to the single tone case, when there is non zero cavity occupation ($n_{\text{eff}} \neq 0$), the interference between the cavity noise and the transduction of the radiation pressure noise would generate noise squashing and anti-squashing effects in the up and down-converted mechanical sidebands.

Inferred mechanical spectrum

The inferred mechanical spectrum is given by normalizing the output spectrum by the gain $\frac{\kappa_R}{\kappa}\Gamma_{\text{opt}}$, which is

$$\bar{S}_{xx,\text{tot}}[\omega] = \frac{\bar{S}_{II}[\omega]}{\frac{\kappa_R}{\kappa}\Gamma_{\text{opt}}} = \frac{\kappa \bar{S}_0[\omega]}{\kappa_R \Gamma_{\text{opt}}} + \left(S'_{bb}[\omega + \delta] + S'_{b^\dagger b^\dagger}[\omega - \delta]\right). \tag{4.106}$$

For ideal cavity where $n_c^{\text{th}} = n_R^{\text{th}} = n_{\text{eff}} = 0$, the inferred mechanical spectrum becomes

$$\bar{S}_{xx,\text{tot}}[\omega] = \frac{\kappa}{2\kappa_R \Gamma_{\text{opt}}} + \frac{\Gamma_m}{(\Gamma_m/2)^2 + (\omega + \delta)^2}\left(n_m^{\text{th}} + n_m^{\text{BA}} + \frac{\beta - \alpha}{2}\right)$$
$$+ \frac{\Gamma_m}{(\Gamma_m/2)^2 + (\omega - \delta)^2}\left(n_m^{\text{th}} + n_m^{\text{BA}} + \frac{\beta + \alpha}{2}\right). \tag{4.107}$$

Taking $\alpha = \beta = 1$, the noise spectrum of the mechanical sidebands in (4.107) reduce to (4.99). However, the physical origin of the asymmetry between the mechanical sidebands are completely different. In (4.99), the asymmetry of the mechanical noise spectrum comes from the quantum fluctuation of the mechanical motion. However, in the inferred mechanical spectrum (4.107), the asymmetry orginate from the quantum fluctuation of the cavity photon. As discussed in the previous section for the single drive case, the sideband asymmetry is entirely due to the detector backaction-imprecision correlation, which is simply the noise squashing and anti-squashing effects of the cavity quantum fluctuation.

4.3.2 Balanced two-tone: Backaction evading measurement (BAE)

As discussed in section 2.2 and shown in the previous subsections, Heisenberg uncertainty principle place a fundamental limit to continuous position measurement, one way to surpass this limit is to perform quantum nondemolition (QND) measurement of special observable as introduced in section 2.3. In this subsection, we will describe how to implement QND measurement of a single mechanical quadrature with a cavity electromechanical system [, ,]. To this end, we drive the system with two drive tones with frequency equal to $\omega_c + \Delta \pm \omega_m$ and phase equal to $\mp\phi$. Then, the Hamiltonian of the system is given by setting $G_\pm = Ge^{\mp i\phi}$ and $\delta = 0$ in Eq. (4.19), which yields

$$\hat{H} = -\hbar\Delta\hat{a}^\dagger\hat{a} - \hbar G\hat{X}_\phi\left(\hat{a}^\dagger + \hat{a}\right), \qquad (4.108)$$

where $\hat{X}_\phi = \cos\phi\hat{X}_1 - \sin\phi\hat{X}_2$ is the mechanical quadrature at phase ϕ. In the following, we set $\phi = 0$ without loss of generality. Eq. (4.108) provides the QND interaction for the measurement of the mechanical quadrature X_1. The quantum Langevin equations of the quadratures in the frequency space is given by setting $G_- = G_+ = G$ and $\delta = 0$ in Eqs. (4.43), which yields

$$\begin{aligned}
\chi_c^{-1}[\omega]\,\hat{U}_1[\omega] &= \sqrt{\kappa}\hat{U}_{1,\text{in}}[\omega] - \Delta\hat{U}_2[\omega], \\
\chi_m^{-1}[\omega]\,\hat{X}_2[\omega] &= \sqrt{\Gamma_m}\hat{X}_{2,\text{in}}[\omega] + 2G\hat{U}_1[\omega]. \\
\chi_c^{-1}[\omega]\,\hat{U}_2[\omega] &= \sqrt{\kappa}\hat{U}_{2,\text{in}}[\omega] + \Delta\hat{U}_1[\omega] + 2G\hat{X}_1[\omega], \\
\chi_m^{-1}[\omega]\,\hat{X}_1[\omega] &= \sqrt{\Gamma_m}\hat{X}_{1,\text{in}}[\omega].
\end{aligned} \qquad (4.109)$$

Note that the mechanical quadrature $\hat{X}_1$ is decoupled from the others. The cavity field is only sensitive to the X_1 quadrature and all the measurement backaction is applied to the unmeasured X_2 quadrature. Therefore, the mechanical quadrature $\hat{X}_1$ is a QND observable.

Mechanical quadrature noise spectrum

By solving the coupled quantum Langevin equations (4.109), we can calculate the noise spectra of the mechanical quadratures, which are given by

$$\bar{S}_{X_1}[\omega] = \frac{1}{2}\int\frac{d\omega'}{2\pi}\left\langle\left\{\hat{X}_1[\omega'],\hat{X}_1[\omega]\right\}\right\rangle = \frac{\Gamma_m}{(\Gamma_m/2)^2 + \omega^2}\left(2n_m^{\text{th}} + 1\right), \qquad (4.110)$$

$$\bar{S}_{X_2}[\omega] = \frac{1}{2}\int\frac{d\omega'}{2\pi}\left\langle\left\{\hat{X}_2[\omega'],\hat{X}_2[\omega]\right\}\right\rangle = \frac{\Gamma_m}{(\Gamma_m/2)^2 + \omega^2}\left[2\left(n_m^{\text{th}} + n_{\text{ba}}\right) + 1\right], \qquad (4.111)$$

the mechanical quadrature $\hat{X}_1$ is unaffected by the measurement, all the measurement backaction (both classical and quantum) is applied to the unmeasured quadrature $\hat{X}_2$ with the backaction quanta

$$n_{\text{ba}} = \frac{\Gamma_{\text{opt}}}{\Gamma_m} \left[\left(n_c^{\text{th}} + \frac{1}{2} \right) + \left(\frac{4\Delta}{\kappa} \right)^2 \frac{\Gamma_{\text{opt}}\Gamma_m}{(\Gamma_m/2)^2 + \omega^2} \left(n_m^{\text{th}} + \frac{1}{2} \right) \right], \tag{4.112}$$

where $\Gamma_{\text{opt}} = \frac{1}{1+(2\Delta/\kappa)^2} \frac{4G^2}{\kappa}$.

Transmission spectrum

The transmission spectrum of an electromechanical system is given by setting $G_- = G_+ = G$ and $\delta = 0$ in Eq. (4.32), which yields

$$S_{21} = -\frac{\sqrt{\kappa_R\kappa_L}}{\kappa/2 - i\,(\omega + \Delta)}. \tag{4.113}$$

The interferences between the probe field and motional sidebands of the two drives cancel each other, which doesn't modify the cavity transmission spectrum.

Output noise spectrum

The output noise spectrum is given by setting $G_- = G_+ = G$ and $\delta = 0$ in Eq. (4.39), which yields

$$\bar{S}_{II}[\omega] = \bar{S}_0[\omega] + \frac{\kappa_R}{\kappa}\Gamma_{\text{opt}}\bar{S}_{X_1}[\omega]. \tag{4.114}$$

The output noise spectrum consists of the transduced mechanical quadrature noise spectrum and the noise background

$$\bar{S}_0[\omega] = n_R^{\text{th}} + \frac{1}{2} + \frac{\kappa_R\kappa}{(\kappa/2)^2 + (\omega + \Delta)^2}\left(n_c^{\text{th}} - n_R^{\text{th}} \right). \tag{4.115}$$

Inferred mechanical quadrature spectrum

The inferred mechanical quadrature spectrum is given by normalizing the output noise spectrum by the gain $\frac{\kappa_R}{\kappa}\Gamma_{\text{opt}}$, which is

$$\bar{S}_{xx,\text{tot}}[\omega] = \frac{\bar{S}_{II}[\omega]}{\frac{\kappa_R}{\kappa}\Gamma_{\text{opt}}} = \frac{\kappa\bar{S}_0[\omega]}{\kappa_R\Gamma_{\text{opt}}} + \bar{S}_{X_1}[\omega]. \tag{4.116}$$

Since there is no backaction apply to the X_1 quadrature, one can achieve imprecision below the quantum limit of the position measurement by increasing the gain Γ_{opt} of the detector. Analysis including the bad cavity effect for $\Delta = 0$ is given in [ ], and it contributes an additional backaction

$$n_{bad} = \frac{n_{ba}}{32}\left(\frac{\kappa}{\omega_m}\right)^2 \tag{4.117}$$

in both mechanical quadratures.

4.3.3 Two-tone reservoir engineering: Mechanical squeezing

In the previous section, we have explored the dynamical effects of the radiation pressure in an electromechanical system driven by a single tone: by controlling the detuning of the drive tone, one can manipulate the dynamical radiation pressure to damp or amplify the mechanical motion. We have also shown that by driving the electromechanical system with equal amplitude at $\omega_c \pm \omega_m$, one can cancel the backaction from the two tones and perform a BAE measurement of a single mechanical quadrature. Here we will explore the effects of the dynamical backaction of a cavity electromechanical system under two drive tones by perturbing the BAE measurement setup. We consider driving the cavity with two tones at $\omega_\pm = \omega_c \pm \omega_m$ with different amplitude [ ]. The Hamiltonian of the system is given by setting $\Delta = \delta = 0$ in Eq. (4.19), which yields

$$\hat{H} = -\hbar\left[\left(G_-\hat{b} + G_+\hat{b}^\dagger\right)a^\dagger + \left(G_-^*\hat{b}^\dagger + G_+^*\hat{b}\right)\hat{a}\right]. \tag{4.118}$$

If we define the Bolgoliubov mode of the mechanical oscillator

$$\hat{\beta} = \cosh(r)\,\hat{b} + \sinh(r)\,\hat{b}^\dagger, \tag{4.119}$$

where $r = \tanh^{-1}(G_+/G_-)$ is the squeezed parameter. The Hamiltonian can be expressed as

$$\hat{H} = -\hbar\mathcal{G}\left(\hat{a}^\dagger\hat{\beta} + \hat{a}\hat{\beta}^\dagger\right), \tag{4.120}$$

where the coupling $\mathcal{G} = \sqrt{G_-^2 - G_+^2}$. Here we consider $G_- > G_+$ to ensure stability. The Hamiltonian is in beam-splitter form as in the sideband cooling case. However, instead of cooling the mechanical mode $\hat{b}$, it cools the Bogoliubov-mode $\hat{\beta}$. Because the vacuum of $\hat{\beta}$ is the squeezed state $\hat{S}(r)|0\rangle$ (where $\hat{S}(r) = \exp\left[r\left(\hat{b}\hat{b} + \hat{b}^\dagger\hat{b}^\dagger\right)\right]$). This cooling directly leads to stationary squeezed state with squeezed quadrature variance

$$\langle \hat{X}_1^2 \rangle = e^{-2r}\left(1 + 2\langle \hat{\beta}^\dagger \hat{\beta}\rangle + \langle \hat{\beta}\hat{\beta}\rangle + \langle \hat{\beta}^\dagger \hat{\beta}^\dagger\rangle\right). \tag{4.121}$$

The cavity acts as an engineered reservoir that cools the mechanical oscillator into a squeezed state.

Bolgoliubov mode spectrum

We first examine the Bolgoliubov mode of the mechanical motion. The quantum Langevin equations of the cavity field and the Bolgoliubov mode in the frequency space are

$$\chi_c^{-1}[\omega]\,\hat{a}[\omega] = \sqrt{\kappa}\hat{a}_{\text{in}}[\omega] + i\mathcal{G}\hat{b}[\omega], \tag{4.122}$$

$$\chi_m^{-1}[\omega]\,\hat{\beta}[\omega] = \sqrt{\Gamma_m}\hat{\beta}_{\text{in}}[\omega] + i\mathcal{G}\hat{a}[\omega]. \tag{4.123}$$

Substituting Eq. (4.122) into Eq. (4.123), we can express the Bolgoliubov mode in terms of the input field, which yields

$$\hat{\beta}[\omega] = \sqrt{\Gamma_m}\bar{\chi}_m[\omega]\,\hat{\beta}_{\text{in}}[\omega] + i\mathcal{G}\sqrt{\kappa}\chi_c[\omega]\,\bar{\chi}_m[\omega]\,\hat{a}_{\text{in}}[\omega]. \tag{4.124}$$

The Bolgoliubov mode spectrum is given by

$$\bar{S}_{\beta\beta}[\omega] = \int_{-\infty}^{\infty}\frac{d\omega'}{2\pi}\langle\hat{\beta}^\dagger[\omega']\,\hat{\beta}[\omega]\rangle = 4\frac{\left(\kappa^2 + 4\omega^2\right)\Gamma_m n_\beta^{\text{th}} + 4\mathcal{G}^2\kappa n_c^{\text{th}}}{\left(4\mathcal{G}^2 + \Gamma_m\kappa\right)^2 + 4\left(\Gamma_m^2 + \kappa^2 - 8\mathcal{G}^2\right)\omega^2 + 16\omega^4}, \tag{4.125}$$

where $n_\beta^{\text{th}} = \left(\frac{G_-}{\mathcal{G}}\right)^2 n_m^{\text{th}} + \left(\frac{G_+}{\mathcal{G}}\right)^2\left(n_m^{\text{th}} + 1\right)$ is the Bolgoliubov mode occupation of the phonon bath. The Bolgoliubov mode occupation is given by the integral of the spectrum, which gives

$$\bar{n}_\beta = \int_{-\infty}^{\infty} \frac{d\omega}{2\pi} S_{\beta\beta}[\omega] = \frac{\Gamma_m\,(\Gamma_{\text{tot}} + \kappa)}{\Gamma_{\text{tot}}\,(\kappa + \Gamma_m)} n_\beta^{\text{th}} + \frac{\kappa\Gamma_{\text{opt}}}{\Gamma_{\text{tot}}\,(\kappa + \Gamma_m)} n_c^{\text{th}}, \tag{4.126}$$

where $\Gamma_{\text{tot}} = \Gamma_m + \Gamma_{\text{opt}}$ and $\Gamma_{\text{opt}} = 4\mathcal{G}^2/\kappa$.

In weak coupling regime ($\kappa \gg \mathcal{G}$), the Bolgoliubov mode spectrum can be simplified to

$$S_{\beta\beta}[\omega] = \frac{\Gamma_{\text{tot}}}{(\Gamma_{\text{tot}}/2)^2 + \omega^2}\bar{n}_\beta. \tag{4.127}$$

Similar to sideband cooling with a single red-detuned drive, the Bolgoliubov mode occupation can be expressed by the detailed balance equation

$$\bar{n}_\beta = \frac{\Gamma_m}{\Gamma_m + \Gamma_{\text{opt}}} n_\beta^{\text{th}} + \frac{\Gamma_{\text{opt}}}{\Gamma_m + \Gamma_{\text{opt}}} n_c^{\text{th}}. \tag{4.128}$$

In this case, the driven cavity serves as a cold reservoir to the Bolgoliubov mode. The dynamical backaction induced by the two tones dissipatively drives the mechanics into a stationary squeezed state.

Mechanical quadrature spectrum

We can also examine the mechanical squeezing by calculating the mechanical quadrature spectra. The quantum Langevin equations in the frequency space are given by setting $\Delta = \delta = 0$ in Eqs. 4.43, which yields

$$\chi_c^{-1}[\omega]\,\hat{U}_1[\omega] = \sqrt{\kappa}\hat{U}_{1,\text{in}}[\omega] - \bar{G}\hat{X}_2[\omega], \tag{4.129}$$

$$\chi_m^{-1}[\omega]\,\hat{X}_2[\omega] = \sqrt{\Gamma_m}\hat{X}_{2,\text{in}}[\omega] + \tilde{G}\hat{U}_1[\omega], \tag{4.130}$$

$$\chi_c^{-1}[\omega]\,\hat{U}_2[\omega] = \sqrt{\kappa}\hat{U}_{2,\text{in}}[\omega] + \tilde{G}\hat{X}_1[\omega], \tag{4.131}$$

$$\chi_m^{-1}[\omega]\,\hat{X}_1[\omega] = \sqrt{\Gamma_m}\hat{X}_{1,\text{in}}[\omega] - \bar{G}\hat{U}_2[\omega]. \tag{4.132}$$

There are two decoupled set of equations. Compare to the quantum Langevin equations in the BAE measurement (Eqs. (4.109)), the $\hat{X}_1$ quadrature is not intact due to the imbalanced drive ($\bar{G} = G_- - G_+ > 0$). The cavity quadratures $\hat{U}_{1,2}$ measure and perturb the mechanical quadratures $\hat{X}_{2,1}$, which can be viewed as a coherent feedback process. We will show that

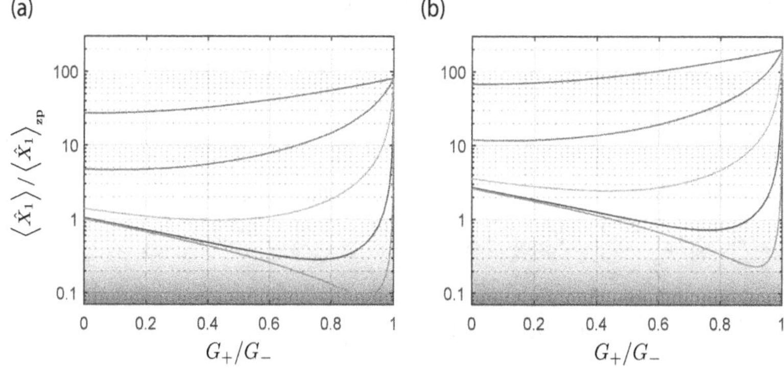

Figure 4.2: The squeezed quadrature occupation (4.135) with different drive ratio G_+/G_-. Different curves represent different total pump photon number ($n_p^{\text{tot}} = n_p^- + n_p^+$). $n_p^{\text{tot}} = 10^2$ (blue), 10^3 (red), 10^4 (yellow), 10^5 (purple), 10^6 (green). The shade blue region indicates quantum squeezing (i.e. $\langle\hat{X}_1\rangle/\langle\hat{X}_1\rangle_{\text{zp}} < 1$). For both plots, $\Gamma_m = 2\pi \times 10\text{Hz}$, $\kappa = 2\pi \times 330\text{kHz}$, and $g_0 = 2\pi \times 130\text{Hz}$. (a) Ideal cavity and mechanical occupations ($n_c^{\text{th}} = 0, n_m^{\text{th}} = 40$). (b) Includes cavity heating and mechanical heating ($n_c^{\text{th}} = 0.8, n_m^{\text{th}} = 100$).

the dynamical backaction from this coherent feedback leads to the stationary mechanical squeezing.

Solving the quantum Langevin equations, we obtain the solutions of the mechanical quadratures

$$\hat{X}_{1,2}[\omega] = \sqrt{\Gamma_m}\,\bar{\chi}_m[\omega]\,\hat{X}_{1,2,\text{in}}[\omega] - \sqrt{\kappa}\,(G_- \mp G_+)\,\chi_c[\omega]\,\bar{\chi}_m[\omega]\,\hat{U}_{2,1,\text{in}}[\omega]. \quad (4.133)$$

The mechanical quadrature spectra are given by

$$\bar{S}_{X_{1,2}}[\omega] = \frac{1}{2}\int \frac{d\omega'}{2\pi}\left\langle\left\{\hat{X}_{1,2}[\omega'],\hat{X}_{1,2}[\omega]\right\}\right\rangle$$
$$= 4\frac{\left(\kappa^2 + 4\omega^2\right)\Gamma_m\left(2n_m^{\text{th}} + 1\right) + 4\left(G_- \mp G_+\right)^2\kappa\left(2n_c^{\text{th}} + 1\right)}{\left(4G^2 + \Gamma_m\kappa\right)^2 + 4\left(\Gamma_m^2 + \kappa^2 - 8G^2\right)\omega^2 + 16\omega^4}. \quad (4.134)$$

Integrating the quadrature spectra, we obtain the mechanical quadrature variances

$$\langle \hat{X}_{1,2}^2 \rangle = \frac{4G^2 + \kappa(\kappa + \Gamma_m)}{(\kappa + \Gamma_m)(4G^2 + \kappa\Gamma_m)} \Gamma_m \left(2n_m^{\text{th}} + 1\right) + \frac{4(G_- \mp G_+)^2}{(\kappa + \Gamma_m)(4G^2 + \kappa\Gamma_m)} \kappa \left(2n_c^{\text{th}} + 1\right).$$

(4.135)

In weak coupling regime ($\kappa \gg G, \Gamma_m$), the mechanical quadrature spectra and the mechanical quadrature variances can be simplified to

$$\bar{S}_{X_{1,2}}[\omega] = \frac{\Gamma_{\text{tot}}}{(\Gamma_{\text{tot}}/2)^2 + \omega^2} \langle \hat{X}_{1,2}^2 \rangle,$$

(4.136)

$$\langle \hat{X}_{1,2}^2 \rangle = \frac{\Gamma_m}{\Gamma_{\text{tot}}} \left(2n_m^{\text{th}} + 1\right) + \frac{\Gamma_{\text{opt}}^{\mp}}{\Gamma_{\text{tot}}} \left(2n_c^{\text{th}} + 1\right),$$

(4.137)

where $\Gamma_{\text{opt}}^{\mp} = 4(G_- \mp G_+)^2/\kappa$.

To illustrate the squeezing mechanism, let's define $\Gamma_{\text{opt},\pm} = 4G_\pm^2/\kappa$, then the mechanical quadrature variances can be expressed as

$$\langle \hat{X}_{1,2}^2 \rangle = (2\bar{n}_m + 1) \mp 2 \frac{\sqrt{\Gamma_{\text{opt},-}\Gamma_{\text{opt},+}}}{\Gamma_{\text{tot}}} \left(2n_c^{\text{th}} + 1\right).$$

(4.138)

The first term in Eq. (4.138) represents the damping of both mechanical quadratures due to the excess power of the red-detuned drive, the average mechanical occupation is given by

$$\bar{n}_m = \frac{\Gamma_m}{\Gamma_{\text{tot}}} n_m^{\text{th}} + \frac{\Gamma_{\text{opt},-}}{\Gamma_{\text{tot}}} n_c^{\text{th}} + \frac{\Gamma_{\text{opt},+}}{\Gamma_{\text{tot}}} \left(n_c^{\text{th}} + 1\right),$$

(4.139)

The second term in Eq. (4.138) comes from the interference of the coherent feedback forces generated by two drives (the second term in the right hand side of Eqs. (4.130) and (4.132)), which produce negative correlation in the X_1 quadrature and positive correlation in the X_2 quadrature. Different from the standard parametric squeezing scheme [ ] which damps the fluctuation in one of the quadrature and amplifiers the other one. The two-tone reservoir engineering scheme damp both of the quadratures, and squeeze the mechanical motion by producing negative correlation in the squeezed quadrature.

To quantify the amount of squeezing, we parametrize the enhanced electromechanical coupling rate $G_\pm = \frac{G}{2}(1 \mp f)$, where $f \in [0, 1)$. The quadrature occupations can be written as

$$\langle \hat{X}_1^2 \rangle = \frac{1}{1+fC}\left(2n_m^{\text{th}}+1\right) + \frac{fC}{1+fC}f\left(2n_c^{\text{th}}+1\right),$$ (4.140)

$$\langle \hat{X}_2^2 \rangle = \frac{1}{1+fC}\left(2n_m^{\text{th}}+1\right) + \frac{C}{1+fC}\left(2n_c^{\text{th}}+1\right),$$ (4.141)

where $C = 4G^2/\kappa\Gamma_m$ is the cooperativity. Consider $fC \gg 1$, then

$$\langle \hat{X}_1^2 \rangle \to \frac{1}{fC}\left(2n_m^{\text{th}}+1\right) + f\left(2n_c^{\text{th}}+1\right),$$ (4.142)

$$\langle \hat{X}_2^2 \rangle \to \frac{1}{fC}\left(2n_m^{\text{th}}+1\right) + \frac{1}{f}\left(2n_c^{\text{th}}+1\right).$$ (4.143)

As shown in (4.142), one can make the contribution $2n_c^{\text{th}}+1$ negligible by choosing a small enough f. For any f, one can always increase the cooperativity C to damp the contribution $2n_m^{\text{th}}+1$. Therefore, in principle, this scheme can generate arbitrary large mechanical squeezing. However, in reality, technical problems such as baths heating and mechanical parametric instability [ ] induced by thermal effects [ ] or nonlinearities [ ] limit the accessibility of the parameters f and C. Nevertheless, as shown in Fig. 4.2, this scheme can squeeze the mechanical quadrature below the zero-point level with reasonable device parameters.

Driven response

The driven response is given by setting $\delta = \Delta = 0$ in Eq. (4.32)

$$S_{21}[\omega] = -\frac{\sqrt{\kappa_R \kappa_L}}{\kappa/2 - i\omega + \frac{G^2}{\Gamma_m/2 - i\omega}},$$ (4.144)

which is in the same form as Eq. 4.59. Similar to sideband cooling, in weak coupling regime, the interference between the mechanical sidebands and the probe field generates a Lorentzian dip with linewidth equal to the mechanical damping rate at $\omega = 0$.

Output spectrum

The output spectrum is given by setting $\delta = \Delta = 0$ in Eq. (4.39), which gives

$$\bar{S}_{II}[\omega] = n_R^{\text{th}} + \frac{1}{2} + 4\kappa_R \frac{\kappa\left(\Gamma_m^2 + 4\omega^2\right)\left(n_c^{\text{th}} - n_R^{\text{th}}\right) + 4\Gamma_m \mathcal{G}^2\left(n_\beta^{\text{th}} - n_R^{\text{th}}\right)}{\left(4\mathcal{G}^2 + \Gamma_m\kappa\right)^2 + 4\left(\Gamma_m^2 + \kappa^2 - 8\mathcal{G}^2\right)\omega^2 + 16\omega^4}. \tag{4.145}$$

In weak coupling regime, it can be simplified to

$$\bar{S}_{II}[\omega] = \bar{S}_0[\omega] + \frac{\kappa_R}{\kappa}\Gamma_{\text{opt}} S'_{\beta\beta}[\omega], \tag{4.146}$$

with

$$S'_{\beta\beta}[\omega] = \frac{\Gamma_{\text{tot}}}{\Gamma_{\text{tot}}^2/4 + \omega^2}\left(\bar{n}_\beta - n_{\text{eff}}\right), \tag{4.147}$$

which is in the same form as the cooling spectrum (4.62). However, the mechanical sideband now represents the Bolgoliubov mode. Similar to the cooling case, the nonzero cavity fluctuation generates noise squashing effect which reduces the area of the mechanical sidebands spectrum by n_{eff}.

Chapter 5

Device fabrication and Measurement

This chapter discusses the devices and the measurement setup of the experiment. Starting with the discussion of the design and the fabrication processes of the cavity electromechanical system. A brief review of the devices developed by the Schwab group will be presented, the problems and difficulties of different devices will be discussed. The detailed fabrication procedures of the devices will be given. Then, the measurement techniques and cryogenic setup in the experiments will be discussed.

5.1 Device design

The Schwab group has been developed superconducting electromechanical system for years. The first generation of superconducting electromechanical device was developed when the Schwab group was back to Maryland and Cornell [ ]. In this device, the mechanical element is a doubly-clamped silicon-nitride (SiN) beam. The beam is covered by aluminum thin film to serve as a electrode, capacitively couple to a superconducting half-wave resonator. Fig. 5.1 is a SEM micrograph of the device. With this type of devices, they have shown the seminal works of sideband cooling [ ] and back-action evasion measurements [ ] close to the quantum regime.

Two technical issues that prevent these devices reaching the quantum regime. The first issue is that the electromechanical coupling for this type of device is relatively weak. The

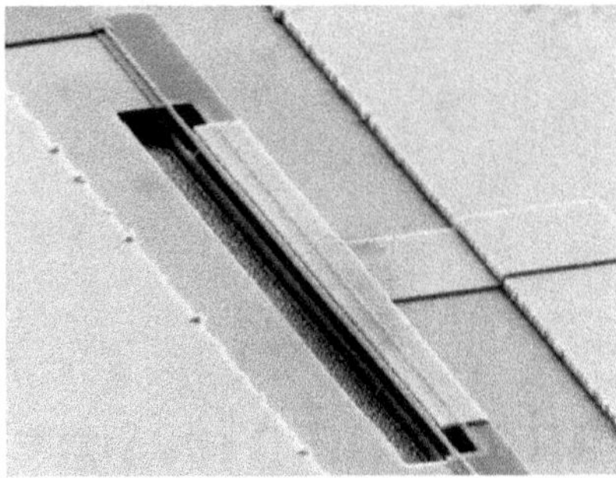

Figure 5.1: Superconducting electromechanical device developed by the Schwab group at Maryland. The mechanical element is the metalized SiN nanobeam (red).

electromechanical coupling rate of this type of system is given by

$$g_0 = -\frac{\partial \omega_c}{\partial x} x_{\text{zp}} = \frac{\omega_c}{2} \frac{1}{C} \frac{\partial C_g}{\partial x} x_{\text{zp}} \sim \frac{1}{2} \eta \frac{\omega_c}{d} x_{\text{zp}}, \tag{5.1}$$

where C_g is the portion of the capacitance that is modulated by the motion of the beam, $\eta = C_g/C$ is the participation factor. In the last step, we make the assumption that $\partial C_g/\partial x \sim C_g/d$. For the nanobeam device, the participation factor is very small ($\eta \sim 1e{-}4$), which ends up with a small electromechanical coupling rate $g_0 \sim 2\pi \times 1$ Hz. In order to achieve large cooperativity factor $C = \frac{4g_0^2}{\kappa \Gamma_m} n_p$, one needs to drive the device with very strong drive tones, which heat up the thermal bath of the mechanical oscillator and the microwave cavity and limit the minimum mechanical occupation in the sideband cooling [ ].

Another issue of the nanobeam devices is the power dependent frequency shift of the mechanical oscillator induced by the 2nd order coupling. If we consider the dependence of the cavity resonance frequency in x into second order, the 2nd order term generates a power dependent spring shift $\Delta \omega_m = \frac{\omega_m}{2} \frac{k_{\text{EM}}}{k}$, where k is the bare mechanical spring constant and the induced electromagnetic spring constant is

$$k_{\text{EM}} = \frac{\hbar \omega_c}{2} |\alpha(t)|^2 \frac{\partial^2 C}{\partial x^2}. \tag{5.2}$$

In the BAE measurement where the cavity power is beating at twice the mechanical resonance frequency, this power dependent frequency shift would generate mechanical parametric amplification. When the mechanical frequency modulation $\Delta\omega_m$ equal to the mechanical linewidth γ_m, the mechanical oscillator start to self-oscillate and become unstable. Consider the ratio

$$\frac{\Delta\omega_m}{\gamma_m} = \frac{Q_m}{2}\frac{k_{\mathrm{EM}}}{k} \sim \left(\frac{\Gamma_{\mathrm{opt}}}{\omega_m}\right)\left(\frac{Q_m}{Q_c}\right)\left(\frac{1}{\eta}\right). \tag{5.3}$$

Since η is very small for nanobeam device, which limit the miximum $\left(\Gamma_{\mathrm{opt}}\right)_{\mathrm{max}} \sim 10$ kHz.

Both of the problems are due to the small participation ratio η. To increase the participation ratio, one need to increase the modulated capacitance C_g, which can be realized by replacing the electrode from 1D nanobeam structure by 2D planar structure [ ]. The work in the Schwab group at Caltech focused on the development of lumped element microwave resonator with vacuum gap planar capacitor and lumped element spiral inductor, the top gate of the capacitor is a suspended metal membrane, which is the mechanical element of the electromechanical system. Compare to the nanobeam design, the membrane design has much higher stiffness and modulated capacitance. For example, for a 5 GHz lumped element resonator with $40 \times 40\ \mu m^2$ paralleled plate capacitor with 100 nm vacuum gap. The modulated capacitance $C_g \simeq 30$ fF, which accounts for 80% of the total capacitance, i.e. $\eta \simeq 0.8$. This design mitigates both of the difficulties in the nanobeam devices for cooling and BAE measurement. In the following, we will discussed the development and fabrication of this type of devices.

5.2 Device fabrication

This section discuss focuses on the fabrication process of the planar electromechanical device developed in the Schwab group. The difficulties through the development of the devices are discussed. The detailed fabrication procedure of the devices in this work are given.

The planar electromechanical devices consist of a spiral inductor shunted by a vacuum gap paralleled plate capacitor. Fig. 5.3 and Fig. 5.4 are some of the devices developed by the Schwab group. In order to fabricate the suspended structure such as the top gate of the capacitor and the bridges, a three layer process is used. Fig. 5.2 outlines the general

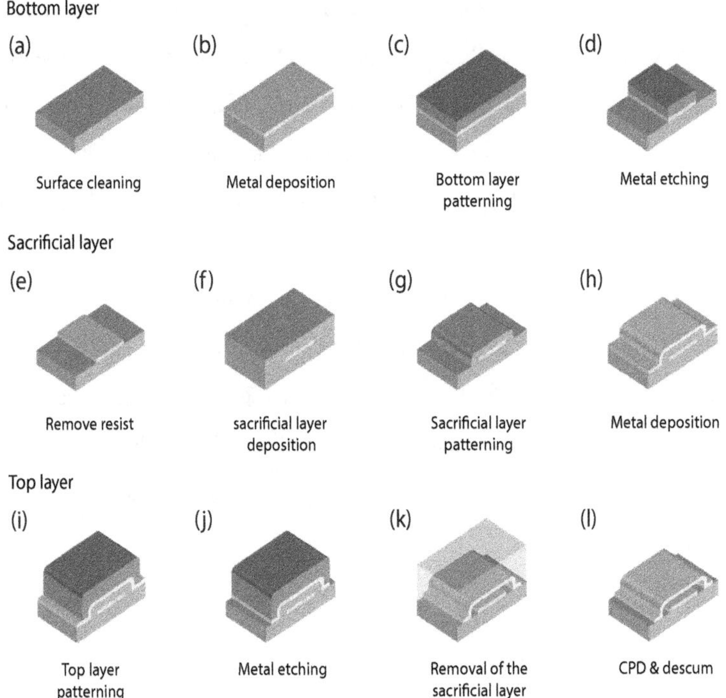

Figure 5.2: Outline of device fabrication processes. Blue, gray, brown, and green colors represent the substrate, metal, resist, and the sacrificial layer respectively.

fabrication procedures. We start with cleaning the the substrate with a modified RCA clean and BOE oxide stripe (Fig. 5.2a). Then, a metal layer is deposited on the substrate (Fig. 5.2b). The bottom layer is patterned with standard optical lithography (Fig. 5.2c) and the metal is removed by either wet etching or dry etching (Fig. 5.2d). After cleaning off the photoresist (Fig. 5.2e), a layer of sacrificial material is deposited on top of the bottom layer (Fig. 5.2f). Then, the sacrificial layer is being patterned (Fig. 5.2g) and a layer metal is deposited on top of it (Fig. 5.2h). The top layer is patterned (Fig. 5.2i) and etched (Fig. 5.2j). The sacrificial layer is then removed with solvent or plasma depending on the material of the sacrificial layer (Fig. 5.2k). Finally, the device is dry with the critical point dryer if the sacrificial layer is removed by solvent, then cleaned with O_2 plasma.

5.2.1 Previous works in the Schwab group

Our group has developed several electromechanical devices with different materials. The development of the device is a long journey, and the results are the efforts of several postdocs and graduate students. A brief review of the device development is given here. Detailed discussion can be found in [   ].

The first generation of device in the Schwab group at Caltech is made of sputtered NbTiN on high resist silicon substrate (Fig. 5.3a) which is aiming for high microwave Q and reduce the nonlinearities associated with two-level system defects [ ]. The sacrificial layer is sputtered SiO_2, which is removed in a buffered-oxide etch (BOE) release soak. However, this device showed low microwave Q and significant Ohmic heating at the pump powers required for sufficient electromechanical coupling. Moreover, the mechanical mode has a small enough specific heat and fast enough thermal time constant for the phonon bath to follow the modulation of cavity power, which generates thermal-induced mechanical parametric instability at the required power in the BAE measurement and limits the measurement imprecision to $2.5x_{zp}$ [   ].

In order to reduce the Ohmic heating and the mechanical bath heating, the next device utilize thermal-evaporated aluminum and SiN membrane as the mechanical element due to the group's past experience (Fig. 5.3b). The sacrificial layer of this device is made out of

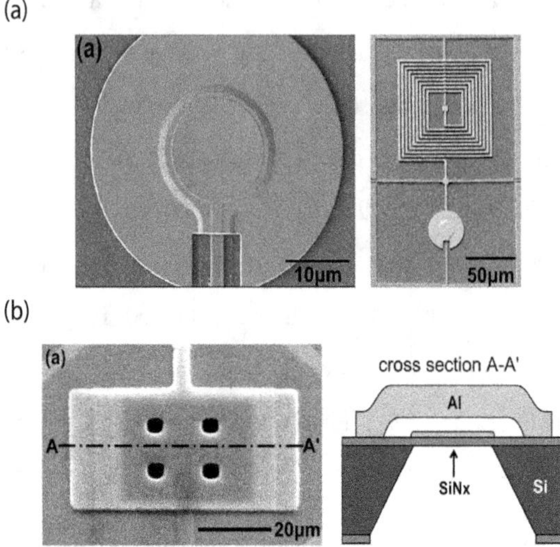

Figure 5.3: Superconducting devices developed by the Schwab group at Caltech. (a) NbTiN device. (b) Al device on SiN.

e-baem evaporated germanium, which can be removed by dry XeF_2 etch [ ]. Although no Ohmic heating and the mechanical bath heating in this device, it showed significant TLS nonlinearities. The TLS nonlinearities induce mechanical parametric instability and limit the precision of the BAE measurement to $1.4x_{zp}$. This TLS defects may be due to the choice of sacrificial layer (germanium alloys with aluminum) and the sacrificial etch (XeF_2 slowly attacks nitride).

5.2.2 Polymer sacrificial layer with sputtered aluminum

With the experiences from the previous works that the choice of the sacrificial and the etching process of it seems to be critical to the quality of the device, we next focus on developing a sacrificial layer that interacts weakly with the metal and the substrate, and requires gentle etching to be removed. Polymer such as e-beam resist or photoresist satisfies both of the requirement. For the next device, we use PMGI as a sacrificial layer which can be removed with Remover-PG. E-beam evaporated aluminum is used for better deposition control. The device showed improved TLS behavior. However, the stress in the top gate

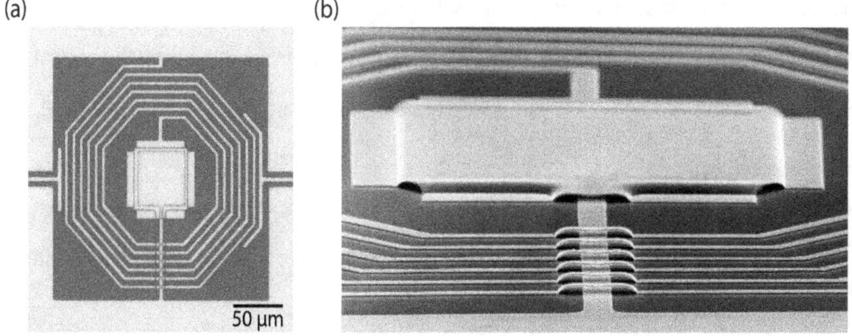

Figure 5.4: (a) Optical micrographs of the device. (b) SEM micrograph of the suspended aluminum membrane.

was difficult to controlled, which make it difficult to make a small gap.

To control the stress of the metal thin film of the top gate, replacing the e-beam deposition with sputter deposition will require minimal recipe modification. However, the sputtered aluminum film has three issues: first, the sputtered aluminum film is not completely removed with transene etchant. It was left with sparkled aluminum oxide residue which was confirmed by energy-dispersive X-ray spectroscopy (EDS). Second, sputtered aluminum film form hillock after annealed above 130°, which would perturb the behavior of the gap. Third, the sputtered aluminum has very poor adhesion with the PMGI even after roughening up the PMGI surface with reactive-ion etch, which is not compatible to the aluminum wet etching process.

The first issue can be solved by a second aluminum etch with low dosage of TMAH, which is common to photoresist developers (e.g., CD-26, MF-319), which can completely remove the residue. The second and the third issues require another polymer sacrificial layer which can be processed with lower bake temperature ($< 130\,°C$) and has good adhesion with sputtered aluminum. We then tried Microposit S1800 series photoresist for the sacrificial layer which require only $115\,°C$. We started with S1813, which can be spined down to $1.5\,\mu m$. In order to obtain reasonable electromechanical coupling rate, we need to thin down the sacrificial layer, which can be realized by a weak flood-exposure with a calibrated exposure time ~ 0.1" before developing the photoresist. Followed by a short (~ 20") RIE with O_2 plasmma to roughen the sacrificial layer, the resulting polymer sacrificial layer has good adhesion with the sputtered aluminum. With this technique, we can control the gap to ~ 100 nm with careful calibrate the duration of the double exposure. Fig. 5.4 are the optical and SEM micrograph of the device with sputtered aluminum and S1813 sacrificial layer. We also tried a thinner photoresist S1805, which can be spin coated to ~ 500 nm, aiming for precise control of the thickness. However, the S1805 was hardened by the processes and couldn't be removed to release the top gate. A detailed fabrication recipe is given in appendix C.

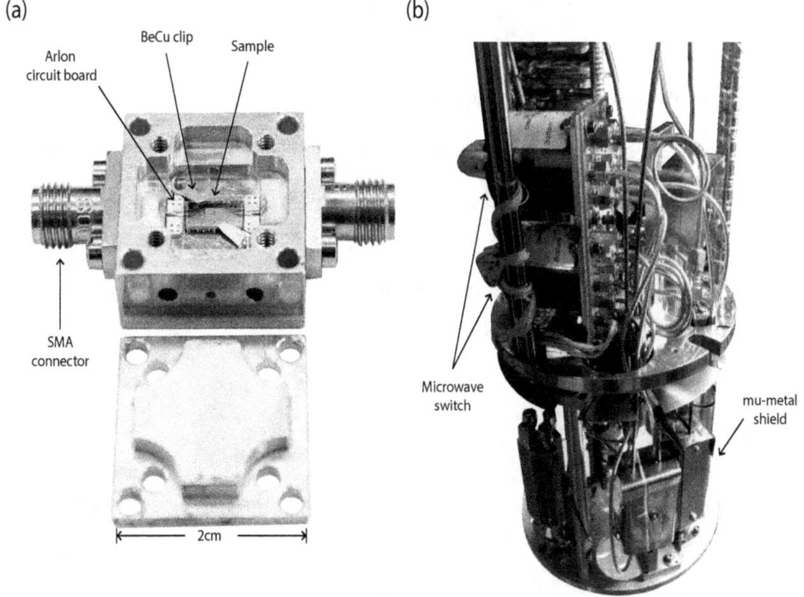

Figure 5.5: (a) Photograph of a device mounted on a sample package. (b) Mixing plate of the dilution refrigerator. The sample packages are mounted on the mixing plate and shielded with mu-metal shields. The two Radiall microwave switches (R573423600) enable us to measure up to six devices in the same cool down.

5.3 Measurement setup

5.3.1 Device packaging

The devices are processed on 6mm × 3mm silicon chips. Which are mounted in a old-plated copper sample package using beryllium copper clips (Fig. 5.3.1a), a small amount of PMMA photoresist is applied to the bottom of the chip for additional thermalization. The ground plane of the chip is wirebounded to the sample package with aluminum wire. The chip waveguides are wirebounded to the ground coplanar waveguide transmission lines on the arlon PCB which is impedance matched to the silicon silicon wafer. The arlon circuit boards are indium soldered to the sample package and the center pins of the PCB transmission lines are soft soldered to the center pin of the SMA connectors. Southwest Microwave SMA connector (214-5 series) is used convert the signal from the coaxial transmission lines to the grounded coplanar waveguides. We have also tested the Non-magnetic version of the SMA connector from Southwest Microwave, but no apparent improvement of the microwave Q was observed.

The sample package is mounted on a cold finger in the mixing plate of the dilution fridge and surrounded by a mu-metal shield (Fig. 5.3.1b). A cryoperm radiation shield is mounted on the mixing plate to provide additional magnetic shielding. We have explored different magnetic shielding configuration [ ], but no obvious improvement in performance is observed.

5.3.2   Fridge circuit

All measurements in this work are performed on the Oxford Kelvinox 400 dilution refrigerator. The fridge is mounted to a floating optical table supported on sand-filled stainless steel pillars for vibration isolation, and is located within a shielding room to minimize electrical noise. Low-temperature thermometry is provided by calibrated RuO thermometers supplied by Oxford which is reliable down to 20 mK. an AVS-47B resistance bridge is used to read out the temperature. A Picowatt TS-530A temperature controller is used to control the temperature of the fridge through a heater at the mixing chamber stage.

In order to probe the device with microwave, we need to connect the device on the mixing plate of the fridge to room temperature with microwave cables. Therefore, the microwave cables in the fridge need to have very low thermal conductivity to thermally isolate the fridge to room temperature. In addition to conduction, the Johnson noise from dissipative elements at room temperature can heat up the fridge through the microwave cables, which is required to either be blocked or attenuated. Fig. 5.6 shows the two fridge wiring configurations in the measurement of device 1 (Fig. 5.6a) and device 2 (Fig. 5.6b). In both configurations, the input line is mainly composed of CuNi coaxial cables (blue lines in Fig. 5.6), which provide similar thermal conductivity and better microwave transmission as compared to stainless-steel coaxial cable. The output line is mainly composed of superconducting Nb coaxial cables (red line in Fig. 5.6) to reduce cable losses.

In the measurement of device 1, cryogenic attenuators with 10, 5, 8, 16 dB attenuation

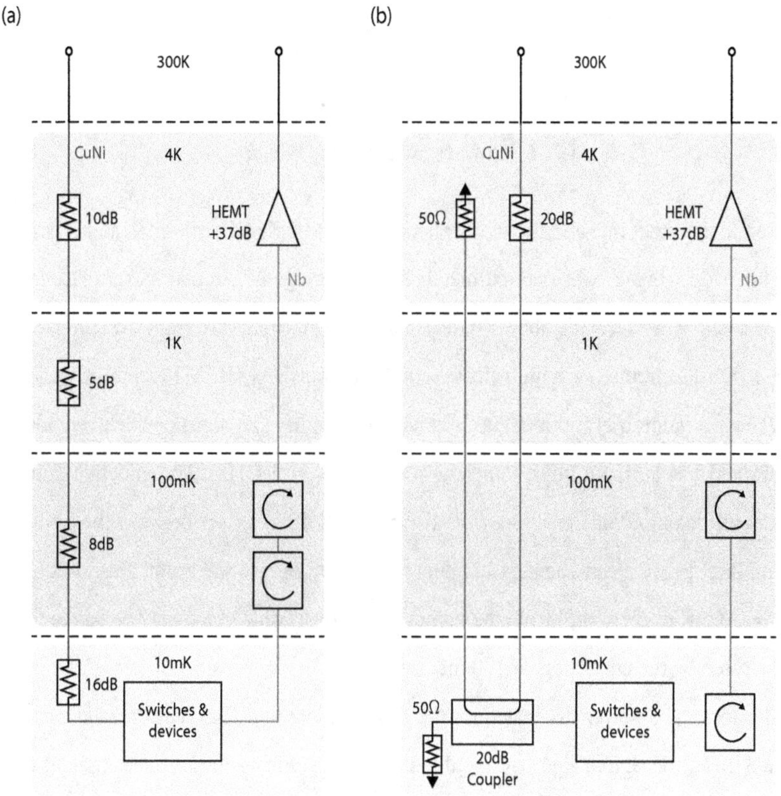

Figure 5.6: Comparison of the fridge circuits for device 1 and 2. The blue lines and the red lines represent the CuNi cables and the Nb cables. (a) Fridge circuit for device 1. The attenuators with 39 dB attenuation are placed at different temperature stages to provide sufficient noise reduction to attenuate the Johnson noise from room temperature. Two cryogenic circulators are placed at the 100 mK stage to block the 4 K noise from the HEMT amplifier. (b) Fridge circuit for device 1. The attenuation is provided by a single 20 dB attenuator at the 4 K stage and a directional coupler at the mixing plate. One of the circulators is thermally anchored to the mixing plate of the fridge.

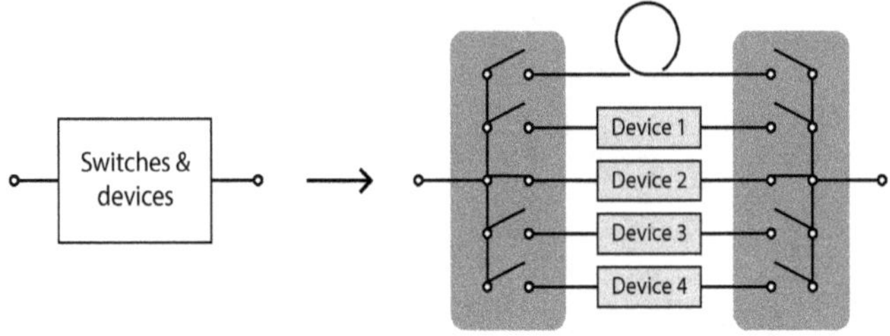

Figure 5.7: Microwave switches and devices.

are installed in the input line and thermally anchored at the 4 K, 1 K, 100 mK, and 10 mK stages of the fridge (Fig 5.6a). Two mu-metal shielded cryogenic circulators are placed at the 100mK stage, which provide about 40 dB isolation to the 4 K noise from the cryogenic amplifier. After the circulators, a high electron mobility transistor (HEMT) amplifier (CIT-4254-077) from Weinreb group at Caltech is used to amplify the signal, which provides 38 dB gain and has an input noise temperature of 3.5 K at 5 GHz. The effective noise temperature of the output line is 5.42 ± 0.03 K due to ~ 2 dB of losses between the device and the amplifier [ ]. From the measurement of the cavity noise spectrum, an additional noise ($T_R = 50$ mK or $n_R^{\mathrm{th}} \simeq 0.2$) from the output port is observed, which may be due to the power dissipated by the circulators at 100 mK stage.

In order to reduce power dissipated on the mixing plate and eliminate the additional noise power from the output line, we modified the fridge wiring in the measurement of device 2. As shown in Fig. 5.6b, a single cryogenic attenuator is placed at the 4K stage and a 20dB directional coupler is placed on the mixing plate. The transmitted port of the directional coupler is connected to a 50Ω at the 4K stage and the coupled port is connected to the device. Therefore the majority of the power going into the coupler is dissipated at the 4 K stage, which has much higher cooling power. In addition, one of the circulators is moved from 100 mK stage to the mixing plate to thermally anchor it to the base temperature of the fridge. After these modifications, the additional noise is eliminated.

Since it takes days to cool the fridge down to base temperature. To facilitate the measurement, two Radiall microwave switches (R573423600) are mounted on the mixing plate, which enable us to measure up to six devices in the same cool down. To reset and

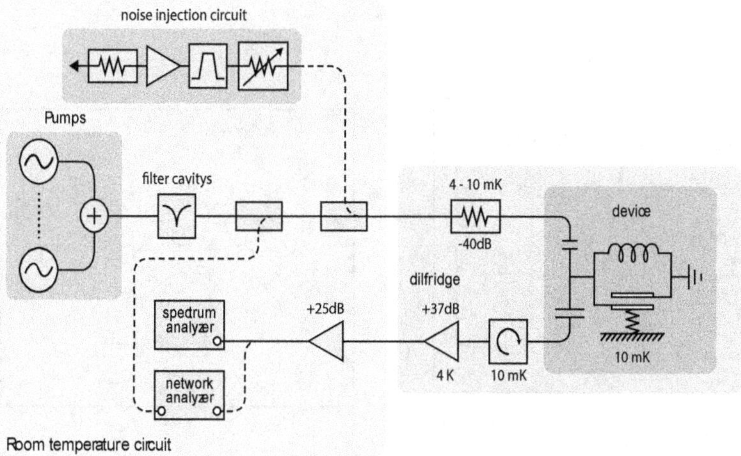

Figure 5.8: Schematic of the measurement circuit. Microwave sources are combined to provide the desired drive signal. Filter cavities are used to filter the excess phase noise of the pumps. The optional noise injection circuit can add classical noise to the device for calibration. The input signal at room temperature is attenuated by 40 dB at different temperature stages in the dilution fridge to remove the room temperature johnson noise. The transmitted signal goes through the circulators and is amplified by a HEMT amplifier, then further amplified by a low noise amplifier at room temperature. The resulting output signal is fed into a spectrum analyzer or a vector network analyzer.

switch to a new connection of the switch, a pulses current is sent to the fridge to drive the actuator. This process would heat the mixing stage up to ~ 100 mK. The fridge then takes about 30 minutes to cool back to the base temperature. Although the switch has six connections, due to the limitation of the available dc lines in the fridge, we only used five of the connections. Cables with the same type and same length as those connecting the device and the switches are installed in the first connection for calibrations. Therefore, we can measure four devices in a single cool down.

5.3.3 Room temperature circuit

Fig. 5.8 shows the simplified schematic diagram of the measurement circuit. Microwave sources are filtered by filter cavities [,] to remove the excess phase noise at room temperature and sent to the fridge from the input port. An optional noise injection circuit

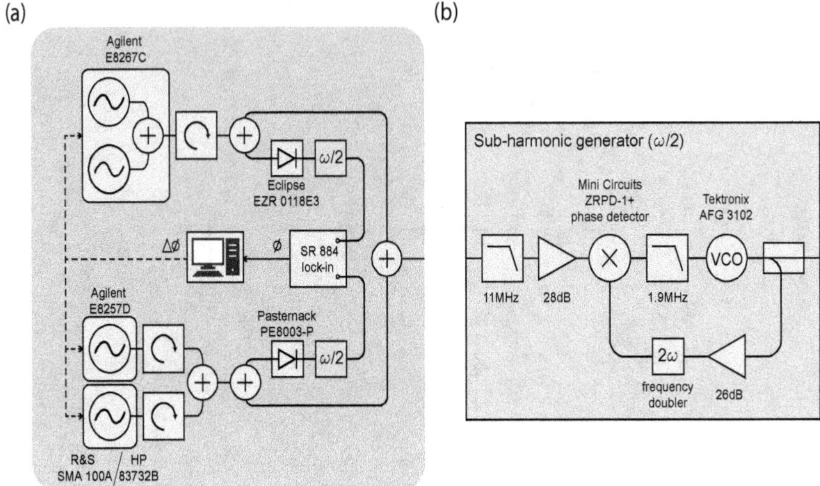

Figure 5.9: Schematic of the drive circuit. (a) The phase locking circuit. (b) The subharmonic circuit.

amplifies the room temperature Johnson noise from a 50 Ω load to generate pseudo-white noise at the resonance of the microwave cavity, which allow us to control the classical noise of the cavity for calibration. The signal from the output port of the fridge is further amplified by 25 dB with a room temperature low noise amplifier (MiTeq LCA 0408) with noise temperature of 120 K. In the experiment, we measure the noise spectrum with a spectrum analyzer (Agilent N9020A) or the driven response with a vector network analyzer (Agilent N5230A or Agilent 8753ES). All sources and measurement devices are synchronized to the 10 MHz rubidium frequency standard (SRS FS725).

Drive circuit

In the double BAE measurement and the BAE measurement of squeezed state, the cavity is driven by two pairs of microwave drive fields. A pair of pump fields

$$\alpha_{\text{pump}} = \left[\alpha_+ e^{-i(\omega_m t + \phi_0)} + \alpha_- e^{i(\omega_m t + \phi_0)} \right] e^{-i\omega_c t} \tag{5.4}$$

are used to prepare the mechanical state. Where the phase ϕ_0 determine a particular direction of the mechanical state (e.g. the direction of the mechanical quadrature $\hat{X}_1$). Another pair of weaker probe fields

$$\alpha_{\text{probe}} = \left[\alpha e^{-i\left(\omega_m t + \phi_{\text{prb}}\right)} + \alpha e^{i\left(\omega_m t + \phi_{\text{prb}}\right)} \right] e^{-i(\omega_c + \Delta)t} \tag{5.5}$$

are used to probe the mechanical state with the BAE measurement. The relative phase $\phi = \phi_{\text{prb}} - \phi_0$ determines the direction of the measured mechanical quadrature $\hat{X}_\phi$ relative to $\hat{X}_1$ with

$$\hat{X}_\phi = \cos\phi \hat{X}_1 - \sin\phi \hat{X}_2. \tag{5.6}$$

The relative phase between the pumps and the probes can be measured by comparing the relative phase of their power beat with a lock-in (Fig. 5.9a). To do that, part of the pumps and the probes are split and fed into power diodes. The output signals from the power diodes are proportional to the power of the pumps and the probes fields, which are

$$|\alpha_{\text{pump}}|^2 = A_+^2 + A_-^2 + 2A_+ A_- \cos\left(2\omega_m t + 2\phi_0\right), \tag{5.7}$$

$$|\alpha_{\text{probe}}|^2 = 2A^2 + 2A^2 \cos\left(2\omega_m t + 2\phi_{\text{prb}}\right). \tag{5.8}$$

The relative phase of the power is equal to 2ϕ, which span from 0 to 4π. Because the lock-in is sensitive to a single branch of phase spanning 2π, in order to cover the full-period in ϕ, the signal from the power diodes are fed into a sub-harmonic circuit (Fig. 5.9b) before compared with the lock-in. The sub-harmonic circuit generate a harmonic signal at ω_m that is phase locked to the diode signal; it uses an arbitrary function generator to produce a reference harmonic signal at ω_m, double it with a frequency doubler, and phase lock it to the diode signal. Then, the relative phase of the resulting signals are compared with a lock-in.

For open loop control, the phase stability is set by the phase drifts of the microwave sources. In our setting, all of the microwave generators are synchronized with a 10 MHz rubidium standard, which provides rms phase drifts on the scale of one degree per 10 minutes. However, our measurements need to average the data for hours. In order to keep the phase drifts small, we separate the spectrum of long average into multiples spectra with a 5 minute average, and adjust the phases for every measurement to keep the phase drift below a degree. In the measurement of device 2, we switch the SMA100A to HP83732B in order to reach the desired frequency range, which worsen the phase stability to one degree per minute. To remedy this issue, we implement a continuous feedback control with a computer to correct the phase per second.

Chapter 6

Backaction evading (BAE) measurement

In this chapter, we will implement a backaction evading (BAE) measurements of the mechanical quadrature with a superconducting electromechanical device. In the first part of the experiment, we will discuss and compare the measurement imprecision and the measurement backaction of a position measurement in balanced detuned two-tone (DTT) configuration and a mechanical quadrature measurement in the two-tone BAE configuration. Using the BAE technique, we realized detection of a single mechanical quadrature with measurement imprecision below the standard quantum limit and reduction of the quantum backaction at the same time. In the second part of the experiment, in addition to the strong BAE pumps, we add another pair of weak BAE probes to detect the measurement backaction induced by the quantum fluctuation of the electromagnetic fields in the microwave resonator. The fridge circuit of the experiments is shown in Fig. 3.6a. The experiments in this chapter are performed in device 1 with the following device parameters:

Parameter	Value	Descriptions
$\omega_m/2\pi$	4.0 MHz	Mechanical resonance frequency
$\Gamma_m/2\pi$	10 Hz	Mechanical linewidth
$\omega_c/2\pi$	5.45 GHz	Cavity resonance frequency
$\kappa/2\pi$	860 kHz	Cavity linewidth
$g_0/2\pi$	15.5 Hz	Electromechanical coupling
x_{zp}	~ 1.8 fm	zero-point motion

Table 6.1: Parameters of device 1.

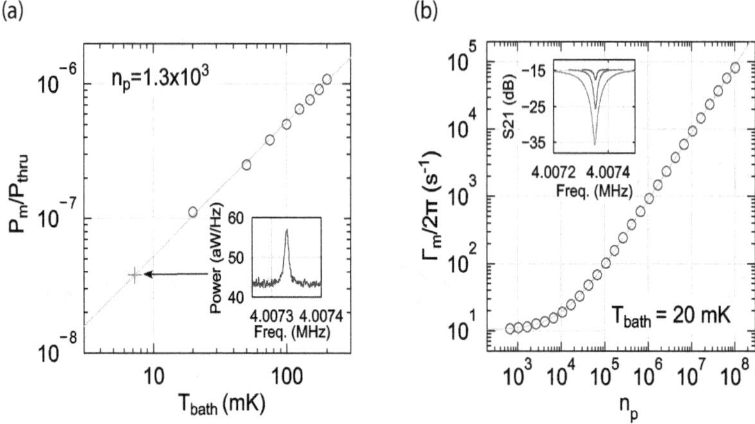

Figure 6.1: Device calibrations. (a) Thermal calibration of the motional sidebands power. The inset is the motional sideband spectrum at base temperature. (b) Linewidth brodening with backaction damping. In addition to a red detuned drive, a weak probe sweeping near the cavity resonance is applied, and its absorption shows the resonant mechanical response. Blue circle, mechanical damping rate; blue line, backaction damping theory fit. The inset is the example of the absorption spectra at $n_p \simeq 5 \times 10^3$, 3×10^4, and 1×10^5 from top to bottom.

6.1 Calibrations

Thermal calibration

In the experiments, we measure the output noise spectrum of the mechanical sideband to quantify the measurement imprecision and the measurement backaction. Therefore, we need to calibrate the noise power of the measured mechanical sideband to the mechanical occupation. To do that, we drive the system with a single red detuned tone at $\omega_- = \omega_c - \omega_m$. We keep the pump power sufficiently low such that the optical damping rate is much smaller than the intrinsic mechanical linewidth ($\Gamma_{opt} \ll \Gamma_m/100$) and the cavity heating effect is negligible ($n_c^{th} \simeq 0$). Then we measure the noise power of the up-converted mechanical sideband P_m and the transmitted power of the red detuned drive P_- at the calibrated temperature, which are given by

$$P_m = \mathcal{G}[\omega_c]\hbar\omega_c\frac{\kappa_R}{\kappa}\Gamma_{\rm opt}n_m^{\rm th}, \tag{6.1}$$

$$P_- = \mathcal{G}[\omega_c]\hbar\omega_p\kappa_R n_p^-, \tag{6.2}$$

where $\Gamma_{\rm opt} = \frac{4g_0^2}{\kappa}n_p^-$ and n_p^- is the intracavity photon number generated by the red detuned tone. $\mathcal{G}[\omega]$ is the gain of the measurement chain which is flat in the measurement bandwidth of the experiments. The gain factor can be eliminated by normalizing the sideband power by transmitted pump power,

$$\frac{P_m}{P_-} = \left(\frac{2g_0}{\kappa}\right)^2 n_m^{\rm th} = \frac{1}{b_-}\frac{k_bT}{\hbar\omega_m}, \tag{6.3}$$

where $b_- = (2g_0/\kappa)^{-2}$ is the calibration factor and k_b is the Boltzmann constant. For this device, the cavity linewidth $\kappa \simeq 869$ kHz is observed to be constant over the relevent pump power. Because the normalized sideband power is linear in the temperature of the mechanical bath, we can extract the calibration factor b_- by measuring the normalized sideband power at different bath temperatures (Fig. 6.1a). From the linear fit, we obtain $b_- = (9.92 \pm 0.16) \times 10^8$ and therefore $g_0 = 2\pi \times 13.8$ Hz, by which we can convert the normalized sideband power into mechanical occupation

$$n_m = b_- \cdot \left(\frac{P_m}{P_-}\right) = (9.92 \pm 0.16) \times 10^8 \cdot \left(\frac{P_m}{P_-}\right). \tag{6.4}$$

Linewidth broadening

In addition to the mechanical occupation, another quantity we want to measure is the intracavity photon number n_p, which can be obtained by the backaction damping measurement. Similar to the thermal calibration, we drive the system with a single red detuned tone at $\omega_- = \omega_c - \omega_m$ to damp the mechanical motion. The optical damping rate is linear to the intracavity photon number and therefore the transmitted pump power

$$\Gamma_{\text{opt}} = \frac{4g_0^2}{\kappa}n_p^- = a_- \left(\frac{4}{\kappa}\right) P_-,\tag{6.5}$$

where $a_- = g_0^2/\mathcal{G}\left[\omega_c\right]\hbar\omega_p\kappa_R$ is the calibration factor. As discussed in section 4.2.1, the interference between the probe field and the red-detuned drive would generate a Lorentzian dip with mechanical linewidth $\Gamma_{\text{tot}} = \Gamma_m + \Gamma_{\text{opt}}$ in the transmission spectrum. By fitting the interference signal in the transmission spectrum (inset in Fig. 6.1b), we can extract the mechanical linewidth Γ_{tot}. Fig. 6.1b shows the total mechanical linewidth with different intracavity photon number. From the fit, we obtain $\Gamma_m = 10$ Hz and the calibration of the intracavity photon number

$$n_p = \left(\frac{a_-}{g_0^2}\right)P_- = (2.25 \pm 0.7) \times 10^{11} \left(\text{W}^{-1}\right) P_-.\tag{6.6}$$

6.2   Backaction evading measurement: Beating the SQL

In this section, we will discuss the experimental results of the backaction evading (BAE) measurement. Using the BAE technique, we realize a mechanical quadrature measurement with measurement imprecision below the standard quantum limit and evading the measurement backaction at the same time.

6.2.1   Definitions of imprecision and backaction

In the experiment, the noise spectrum of the mechanical sideband is measured to quantify the measurement sensitivity. Before discussing the experimental results, here we provide the definitions of the measurement imprecision and the measurement backaction in terms of the measured noise spectra.

6.2.1.1   Quadrature measurement in BAE configuration

We will first discuss the precision of the mechanical quadrature measurement with the BAE technique, in order to compare the precision of the quadrature measurement and the position measurement. In this section, we define the mechanical quadratures with

$$\hat{X}_1 = x_{zp}\left(\hat{b} + \hat{b}^\dagger\right), \quad \hat{X}_2 = -ix_{zp}\left(\hat{b} - \hat{b}^\dagger\right). \tag{6.7}$$

The position of the mechanical oscillator can be expressed as

$$\hat{x} = \hat{X}_1 \cos\left(\omega_m t\right) + \hat{X}_2 \sin\left(\omega_m t\right). \tag{6.8}$$

As discussed in section 4.3.2, in order to perform a BAE measurement of a single mechanical quadrature, the cavity is driven by two drive tones with equal power at $\omega_c \pm \omega_m$ to generate the required modulation of the electromechanical coupling. The up-converted sideband of the red detuned tone and the down-converted sideband of the blue detuned tone are overlapped at the center of the cavity resonance. The measured noise spectrum of the mechanical sideband at the output of the measurement chain is

$$\bar{S}_{\text{out}}^{\text{BAE}}[\omega] = \mathcal{G}[\omega_c]\,\hbar\omega_c\left(\bar{S}_{\text{add}} + \bar{S}_0 + \frac{\kappa_R}{\kappa}\Gamma_{\text{opt}}\frac{\bar{S}_{X_1}[\omega]}{x_{zp}^2}\right), \tag{6.9}$$

where $\bar{S}_0$ is the noise floor of the electromechanical system (Eq. (4.115)), which is flat in the bandwidth of the mechanical sideband spectrum. $\bar{S}_{\text{add}}$ is the added noise of the measurement chain, which is dominated by the add noise of the HEMT amplifier in our experiment. $\bar{S}_{X_1}[\omega]$ is the noise spectrum of the mechanical quadrature (Eq. (4.110)).

Using the thermal calibration, the measured mechanical quadrature spectrum is given by normalizing the output noise spectrum with the transmitted power of the red detuned tone, which gives

$$\frac{\bar{S}_{X_1,\text{tot}}^{\text{BAE}}[\omega]}{x_{zp}^2} = \left(\frac{b_-}{P_-}\right)\bar{S}_{\text{out}}^{\text{BAE}}[\omega] = \frac{\bar{S}_{X_1,\text{imp}}^{\text{BAE}}}{x_{zp}^2} + \frac{\bar{S}_{X_1}[\omega]}{x_{zp}^2}, \tag{6.10}$$

where $\dfrac{\bar{S}_{X_1,\text{imp}}^{\text{BAE}}}{x_{zp}^2} = \dfrac{F b_-}{\kappa_R n_p}$ is the noise floor of the mechanical quadrature measurement and $F = \bar{S}_{\text{add}} + \bar{S}_0$ is the total noise floor of the output spectrum.

The measured mechanical quadrature variance for the BAE measurement is the sum of the measurement imprecision and the mechanical quadrature occupation

$$\frac{\langle \hat{X}_1^2 \rangle_{\text{tot}}^{\text{BAE}}}{x_{\text{zp}}^2} = \frac{\langle \hat{X}_1^2 \rangle_{\text{imp}}^{\text{BAE}}}{x_{\text{zp}}^2} + \frac{\langle \hat{X}_1^2 \rangle}{x_{\text{zp}}^2}, \tag{6.11}$$

which are given by

$$\frac{\langle \hat{X}_1^2 \rangle_{\text{imp}}^{\text{BAE}}}{x_{\text{zp}}^2} = \frac{\bar{S}_{X_1,\text{imp}}^{\text{BAE}}}{x_{\text{zp}}^2} \frac{\Gamma_m}{4} = \left(\frac{\kappa}{\kappa_R} \right) \left(\frac{F}{4} \right) \left(\frac{\Gamma_m}{\Gamma_{\text{opt}}} \right), \tag{6.12}$$

$$\frac{\langle \hat{X}_1^2 \rangle}{x_{\text{zp}}^2} = \frac{\bar{S}_{X_1}[0]\,\Gamma_m}{x_{\text{zp}}^2} \frac{\Gamma_m}{4} = 2n_m^{\text{th}} + 1. \tag{6.13}$$

For ideal BAE measurement, the backaction is entirely routed to the unmeasured mechanical quadrature $\hat{X}_2$ with

$$\frac{\langle \hat{X}_2^2 \rangle}{x_{\text{zp}}^2} = \frac{\bar{S}_{X_2}[0]\,\Gamma_m}{x_{\text{zp}}^2} \frac{\Gamma_m}{4} = 2\left(n_m^{\text{th}} + n_{\text{ba}} \right) + 1, \tag{6.14}$$

where $\bar{S}_{X_2}[\omega]$ is the noise spectrum of the mechanical quadrature $\hat{X}_2$ (Eq. (4.111)) and $n_{\text{ba}} = \frac{\Gamma_{\text{opt}}}{\Gamma_m} \left(n_c^{\text{th}} + \frac{1}{2} \right)$ is the measurement backaction induced by the BAE measurement. No backaction is applied to the measured mechanical quadrature $\hat{X}_1$. As shown in Eq. (6.12), the measurement imprecision is proportional to the inverse of the intracavity pump photon number. Ideally, one can make the measurement imprecision arbitrarily small by increasing the pump power. In practice, cavity and mechanical nonlinearities would pose limits to the applied power, which limits the measurement imprecision above the standard quantum limit in the previous experiments [44, 93, 94].

6.2.1.2 Position measurement in DTT configuration

To quantify the reduction of the backaction in the BAE measurement, we compare the result from the BAE measurement with the result from a position measurement in the detuned two tones (DTT) configuration. As discussed in section 4.3.1, in the DTT configuration, the cavity is driven by two pumps with equal power at $\omega_c \pm (\omega_m + \delta)$. We choose $\kappa \gg \delta \gg \Gamma_m$ such that both of the mechanical sidebands are well separated with each other and sited well inside the cavity resonance. At the output of the measurement chain, the noise spectrum of the two mechanical sidebands are

$$\bar{S}_{\text{out}}^{\text{DTT}}[\omega] = \mathcal{G}[\omega_c]\hbar\omega_c \left\{ \bar{S}_{\text{add}} + \bar{S}_0 + \frac{\kappa_R}{\kappa}\Gamma_{\text{opt}}\frac{S_x^{\text{DTT}}[\omega]}{x_{\text{zp}}^2} \right\}, \tag{6.15}$$

where

$$\frac{S_x^{\text{DTT}}[\omega]}{x_{\text{zp}}^2} = S_{bb}'[\omega + \delta] + S_{b^\dagger b^\dagger}'[\omega - \delta] \tag{6.16}$$

is the noise spectra of the up and down-converted mechanical sidebands, which are given by

$$S_{bb}'[\omega] = \frac{\Gamma_m}{(\Gamma_m/2)^2 + \omega^2}\left(\bar{n}_m^{\text{DTT}} - n_{\text{eff}}\right), \tag{6.17}$$

$$S_{b^\dagger b^\dagger}'[\omega] = \frac{\Gamma_m}{(\Gamma_m/2)^2 + \omega^2}\left(\bar{n}_m^{\text{DTT}} + 1 + n_{\text{eff}}\right), \tag{6.18}$$

where $\bar{n}_m^{\text{DTT}} = n_m^{\text{th}} + n_{\text{ba}}^{\text{DTT}}$ is the mechanical occupation, $n_{\text{ba}}^{\text{DTT}} = \frac{\Gamma_{\text{opt}}}{\Gamma_m}\left(2n_c^{\text{th}} + 1\right)$ is the backaction heating in terms of quanta and $n_{\text{eff}} = 2n_c^{\text{th}} - n_R^{\text{th}}$ is the noise squashing/anti-squashing quanta.

Using the thermal calibration, the measured mechanical spectrum is given by normalizing the output spectrum with the transmitted power of the red detuned drive

$$\frac{S_{x,\text{tot}}^{\text{DTT}}[\omega]}{x_{\text{zp}}^2} = \left(\frac{b_-}{P_-}\right)\bar{S}_{\text{out}}^{\text{DTT}}[\omega] = \frac{\bar{S}_{x,\text{imp}}^{\text{DTT}}}{x_{\text{zp}}^2} + \frac{S_x^{\text{DTT}}[\omega]}{x_{\text{zp}}^2}, \tag{6.19}$$

where $\frac{\bar{S}_{x,\text{imp}}^{\text{DTT}}}{x_{\text{zp}}^2} = \frac{Fb_-}{\kappa_R n_p}$ is the noise floor of the measurement. Note that $S_x^{\text{DTT}}[\omega]$ contains the noise squashing/anti-squashing factor n_{eff}, which can be eliminated by symmetrizing the spectrum

$$\frac{\bar{S}_{x,\text{tot}}^{\text{DTT}}[\omega]}{x_{\text{zp}}^2} = \frac{1}{2}\left(\frac{S_{x,\text{tot}}^{\text{DTT}}[\omega]}{x_{\text{zp}}^2} + \frac{S_{x,\text{tot}}^{\text{DTT}}[-\omega]}{x_{\text{zp}}^2}\right) = \frac{\bar{S}_{x,\text{imp}}^{\text{DTT}}}{x_{\text{zp}}^2} + \frac{\bar{S}_x^{\text{DTT}}[\omega]}{x_{\text{zp}}^2}, \tag{6.20}$$

where

$$\bar{S}_x^{\text{DTT}}[\omega] = \frac{1}{2}\left(S_x^{\text{DTT}}[\omega] + S_x^{\text{DTT}}[-\omega]\right) = \bar{S}_x^0[\omega] + \bar{S}_{x,\text{ba}}^{\text{DTT}}[\omega] \tag{6.21}$$

is the symmetrized noise spectrum of the mechanical motion. The first term is the contribution of the phonon bath, and the second term is the contribution of the measurement backaction, both of which are given by

$$\bar{S}_x^0[\omega] = x_{\mathrm{zp}}^2 \frac{\Gamma_m}{(\Gamma_m/2)^2 + (|\omega| - \delta)^2} \left(n_m^{\mathrm{th}} + \frac{1}{2} \right), \tag{6.22}$$

$$\bar{S}_{x,\mathrm{ba}}^{\mathrm{DTT}}[\omega] = x_{\mathrm{zp}}^2 \frac{\Gamma_m}{(\Gamma_m/2)^2 + (|\omega| - \delta)^2} n_{\mathrm{ba}}^{\mathrm{DTT}}. \tag{6.23}$$

The measured position variance for the DTT measurement is the sum of the measurement imprecision and the fluctuation of the mechanical motion

$$\frac{\langle \hat{x}^2 \rangle_{\mathrm{tot}}^{\mathrm{DTT}}}{x_{\mathrm{zp}}^2} = \left(\frac{\bar{S}_{x,\mathrm{tot}}^{\mathrm{DTT}}[-\delta]}{x_{\mathrm{zp}}^2} + \frac{\bar{S}_{x,\mathrm{tot}}^{\mathrm{DTT}}[\delta]}{x_{\mathrm{zp}}^2} \right) \frac{\Gamma_m}{4} = \frac{\langle \hat{x}^2 \rangle_{\mathrm{imp}}^{\mathrm{DTT}}}{x_{\mathrm{zp}}^2} + \frac{\langle \hat{x}^2 \rangle^{\mathrm{DTT}}}{x_{\mathrm{zp}}^2}. \tag{6.24}$$

The measurement imprecision for the DTT measurement is

$$\frac{\langle \hat{x}^2 \rangle_{\mathrm{imp}}^{\mathrm{DTT}}}{x_{\mathrm{zp}}^2} = 2 \frac{\bar{S}_{x,\mathrm{imp}}^{\mathrm{DTT}}}{x_{\mathrm{zp}}^2} \frac{\Gamma_m}{4} = \left(\frac{\kappa}{\kappa_R} \right) \left(\frac{F}{2} \right) \left(\frac{\Gamma_m}{\Gamma_{\mathrm{opt}}} \right). \tag{6.25}$$

Similar to BAE measurement, the measurement imprecision is proportional to the inverse of the pump power. However, for position measurement, the measurement backaction generates additional fluctuation to the mechanical motion. The total variance of the mechanical motion is the sum of the thermal fluctuation $\langle \hat{x}^2 \rangle_0$ and the measurement backaction $\langle \hat{x}^2 \rangle_{\mathrm{ba}}^{\mathrm{DTT}}$

$$\frac{\langle \hat{x}^2 \rangle^{\mathrm{DTT}}}{x_{\mathrm{zp}}^2} = \left(\frac{\bar{S}_x^{\mathrm{DTT}}[\delta]}{x_{\mathrm{zp}}^2} + \frac{\bar{S}_x^{\mathrm{DTT}}[-\delta]}{x_{\mathrm{zp}}^2} \right) \frac{\Gamma_m}{4} = \frac{\langle \hat{x}^2 \rangle_0}{x_{\mathrm{zp}}^2} + \frac{\langle \hat{x}^2 \rangle_{\mathrm{ba}}^{\mathrm{DTT}}}{x_{\mathrm{zp}}^2}, \tag{6.26}$$

where

$$\frac{\langle \hat{x}^2 \rangle_0}{x_{\mathrm{zp}}^2} = \left(\frac{\bar{S}_x^0[-\delta]}{x_{\mathrm{zp}}^2} + \frac{\bar{S}_x^0[\delta]}{x_{\mathrm{zp}}^2} \right) \frac{\Gamma_m}{4} = 2 n_m^{\mathrm{th}} + 1, \tag{6.27}$$

$$\frac{\langle \hat{x}^2 \rangle_{x,\mathrm{ba}}^{\mathrm{DTT}}}{x_{\mathrm{zp}}^2} = \left(\frac{\bar{S}_{x,\mathrm{ba}}^{\mathrm{DTT}}[-\delta]}{x_{\mathrm{zp}}^2} + \frac{\bar{S}_{x,\mathrm{ba}}^{\mathrm{DTT}}[\delta]}{x_{\mathrm{zp}}^2} \right) \frac{\Gamma_m}{4} = 2 n_{\mathrm{ba}}^{\mathrm{DTT}} = 2 \frac{\Gamma_{\mathrm{opt}}}{\Gamma_m} \left(2 n_c^{\mathrm{th}} + 1 \right). \tag{6.28}$$

As shown in Eq. (6.28), the measurement backaction is proportional to the pump power.

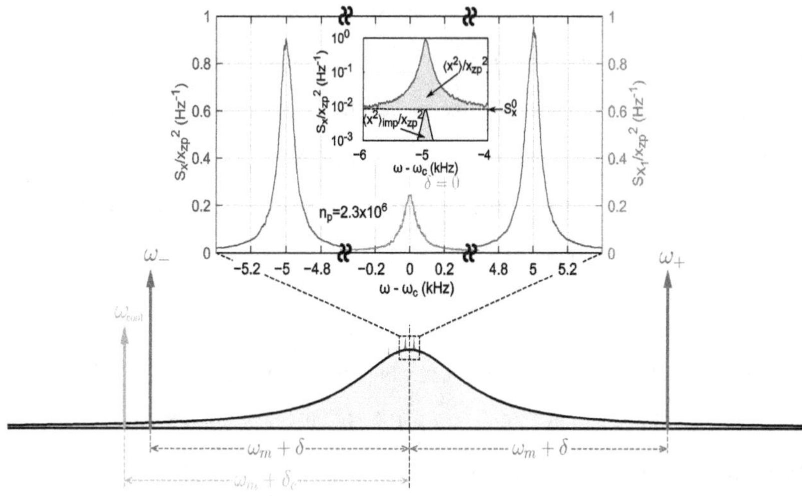

Figure 6.2: Schematic diagram of the pumps configuration. The blue Lorentzian represents the cavity resonance. The green arrow represents the cooling drive. The two purple arrows represent the balanced tones for the position measurement in the DTT configuration ($\delta \gg \Gamma_m$) or BAE measurement of the mechanical quadrature ($\delta = 0$). The inset shows examples of the mechanical spectra in DTT and BAE configurations.

Therefore, one cannot make the total added noise of the position measurement

$$\langle \hat{x}^2 \rangle^{\mathrm{DTT}}_{\mathrm{add}} / x^2_{\mathrm{zp}} = \langle \hat{x}^2 \rangle^{\mathrm{DTT}}_{\mathrm{imp}} / x^2_{\mathrm{zp}} + \langle \hat{x}^2 \rangle^{\mathrm{DTT}}_{\mathrm{ba}} / x^2_{\mathrm{zp}} \tag{6.29}$$

arbitrarily small by ramping up the pump power.

6.2.2 Experimental results

Having analyzed the measurement precision in BAE and DTT configurations, we will now discuss the experimental results. In this experiment, we implement BAE measurement of a single mechanical quadrature with the superconducting electromechanical device. We will quantify and compare the measurement imprecision and the measurement backaction of a single mechanical quadrature measurement with the BAE scheme and the position measurement with the DTT scheme.

Fig. 6.2 shows the pump configuration of the experiment. We drive the cavity with two balanced tones at $\omega_\pm = \omega_c \pm (\omega_m + \delta)$ to perform a single mechanical quadrature measurement in BAE configuration ($\delta = 0$ Hz) or a position measurement in DTT configuration ($\delta = 2\pi \times 5$ Hz). In order to suppress the mechanical frequency jitter, a third red detuned tone is driving the cavity at $\omega_{\text{cool}} = \omega_c - (\omega_m + \delta_c)$ with $\delta_c = 2\pi \times 35$ kHz to broaden the mechanical linewidth. We quantify the measurement imprecision and the measurement backaction with the following protocol:

1. Set the initial mechanical state:

   We first turn on the cooling tone (red arrow in Fig. 6.2) with $n_p^{\text{cool}} \simeq 10^5$ to broaden the mechanical linewidth to $\Gamma_{m,\text{init}} = 2\pi \times 100$ Hz. We measure the noise power of the up-converted mechanical sideband P_m^{cool} and the transmitted pump power P_-^{cool} of the cooling tone to quantify the initial mechanical occupation, which is given by

   $$n_{m,\text{init}}^{\text{th}} = b_- \frac{P_m^{\text{cool}}}{P_-^{\text{cool}}} \simeq 13, \tag{6.30}$$

   which corresponds to $\langle \hat{x}^2 \rangle_{\text{init}} / x_{\text{zp}} = \langle \hat{X}_1^2 \rangle_{\text{init}} / x_{\text{zp}} = 2n_{m,\text{init}}^{\text{th}} + 1 \simeq 27$. The green curve in Fig. 6.3a is the spectrum of the mechanical sideband and the green dots in Fig. 6.3b are the initial mechanical fluctuation in each measurement.

2. Position measurement in DTT configuration:

   Then, we turn on the balanced tones with $\delta = 2\pi \times 5$ kHz and intracavity pump photon number n_p to perform the position measurement in the DTT configuration. We measure the noise spectra of the corresponding mechanical sidebands $\bar{S}_{\text{out}}^{\text{DTT}}[\omega]$ and the transmitted pump power of the red detuned drive P_-. The balancing condition of the two tones is ensured by matching the mechanical linewidth Γ_m to $\Gamma_{m,\text{init}}$ within $2\pi \times 5$ kHz.

   We use the thermal calibration b_- to convert the normalized output noise spectrum to the mechanical noise spectrum $S_{x,\text{tot}}^{\text{DTT}}[\omega]$ with Eq. (6.19), an example of the measured mechanical spectrum is shown in the inset of Fig. 6.2 (blue curves). The measurement imprecision $\langle \hat{x}^2 \rangle_{\text{imp}}^{\text{DTT}} / x_{\text{zp}}^2$ and the mechanical fluctuation $\langle \hat{x}^2 \rangle^{\text{DTT}} / x_{\text{zp}}^2$ in the DTT measurement are given by Eqs. (6.25) and (6.26). The blue curve in

Fig. 6.3a is the symmetrized mechanical spectrum $\bar{S}_{x,\text{tot}}^{\text{DTT}}[\omega]$ around the mechanical resonance. The blue dots in Fig. 6.3b are the mechanical fluctuation in the DTT configuration with different n_p.

3. Mechanical quadrature measurement in BAE configuration:

   To perform BAE measurement of the mechanical quadrature, we overlap the two mechanical sidebands at the center of the cavity resonance ($\delta = 0$ Hz). Similar to the DTT measurement, we measure the output noise spectrum of the overlapped mechanical sideband $\bar{S}_{\text{out}}^{\text{BAE}}[\omega]$ and the transmitted power of the red detuned tone P_-, then convert the output spectrum to the mechanical quadrature spectrum $\bar{S}_{X_1,\text{tot}}^{\text{BAE}}[\omega]$ with Eq. (6.10). Examples of the measured mechanical quadrature spectrum are shown in the inset of Fig. 6.2 (red curve) and Fig. 6.3a (red curve). The measurement imprecision $\langle \hat{X}_1^2 \rangle_{\text{imp}}/x_{\text{zp}}^2$ and the mechanical quadrature fluctuation $\langle \hat{X}_1^2 \rangle/x_{\text{zp}}^2$ are given by Eqs. (6.12) and (6.13). The red dots in Fig. 6.3b are the mechanical quadrature fluctuation measured in the BAE configuration with different n_p.

4. Extract the measurement backaction of the DTT and BAE measurement.

   To extract the backaction from the measurements, we compare the measured mechanical fluctuation of the DTT ($\langle \hat{x}^2 \rangle^{\text{DTT}}/x_{\text{zp}}^2$) and BAE measurements ($\langle \hat{X}_1^2 \rangle/x_{\text{zp}}^2$) to the initial mechanical fluctuation ($\langle \hat{x}^2 \rangle_{\text{init}}/x_{\text{zp}}$ and $\langle \hat{X}^2 \rangle_{1,\text{init}}/x_{\text{zp}}$), which gives

$$\langle \hat{x}^2 \rangle_{\text{ba}} = \langle \hat{x}^2 \rangle^{\text{DTT}} - \langle \hat{x}^2 \rangle_{\text{init}} = \langle \hat{x}^2 \rangle_{\text{ba}}^{\text{diss}} + \langle \hat{x}^2 \rangle_{\text{ba}}^{\text{DTT}}, \tag{6.31}$$

$$\langle \hat{X}_1^2 \rangle_{\text{ba}} = \langle \hat{X}_1^2 \rangle - \langle \hat{X}_1^2 \rangle_{\text{init}} = \langle \hat{X}_1^2 \rangle_{\text{ba}}^{\text{diss}}, \tag{6.32}$$

where $\langle \hat{x}^2 \rangle_{\text{ba}}^{\text{diss}}$ and $\langle \hat{X}_1^2 \rangle_{\text{ba}}^{\text{diss}}$ are the backaction from source other than the the detected field, which cannot be avoided in the BAE measurement. In our experiment, this extra backaction is due to the increase in mechanical fluctuations due to thermal dissipation in the device. The balanced tones heat up the mechanics, and therefore the mechanical fluctuation in the DTT measurement ($n_{m,\text{DDT}}^{\text{th}}$) and BAE measurement ($n_{m,\text{BAE}}^{\text{th}}$) are higher than the initial mechanical fluctuation $n_{m,\text{init}}^{\text{th}}$, which gives the addition measurement backactions

$$\langle \hat{x}^2 \rangle_{\text{ba}}^{\text{diss}} = 2 \left(n_{m,\text{DTT}}^{\text{th}} - n_{m,\text{init}}^{\text{th}} \right) x_{\text{zp}}^2, \tag{6.33}$$

$$\langle \hat{X}_1^2 \rangle_{\text{ba}}^{\text{diss}} = 2 \left(n_{m,\text{BAE}}^{\text{th}} - n_{m,\text{init}}^{\text{th}} \right) x_{\text{zp}}^2. \tag{6.34}$$

Fig. 6.3b shows the increase of the mechanical fluctuation due to the measurement backaction at different n_p.

Fig. 6.3c shows the total measurement backaction, $\langle \hat{x}^2 \rangle_{\text{ba}}$ and $\langle \hat{X}_1^2 \rangle_{\text{ba}}$, and the measurement imprecision, $\langle \hat{x}^2 \rangle_{\text{imp}}$ and $\langle \hat{X}_1^2 \rangle_{\text{imp}}$, in the DTT and BAE configurations at various intracavity pump photon number. In the position measurement with DTT configuration, as the measurement imprecision decreases (blue circles), the measurement backaction increase (blue dots). At $n_p = 2.3 \times 10^6$, the mechanical occupation increases from 13.0 ± 0.5 to 68.5 ± 0.1, consistent with a small finite microwave occupation factor ($n_c \simeq 0.6 \pm 0.1$) in addition to the quantum fluctuations. In contrast, we do not observe a large increase in the mechanical fluctuations in BAE as the imprecision decreases (red dots). The expected backaction into the measured mechanical quadrature due to the finite sideband resolution is $0.12 x_{\text{zp}}^2$ at $n_p = 4.7 \times 10^6$. The measured backaction of $\langle \hat{X}_1^2 \rangle_{\text{ba}} \simeq 10 x_{\text{zp}}^2$ is likely due to thermal dissipation in our device. Nonetheless, we demonstrate avoidance of the backaction noise by 10.7 ± 0.3 dB at $n_p = 2.3 \times 10^6$ compared with the position measurement. Most important, we show that the backaction $\langle \hat{X}_1^2 \rangle_{\text{ba}}$ is 8.5 ± 0.4 dB below the level set by quantum fluctuations of the microwave field, which is $2 \left(\Gamma_{\text{opt}} / \Gamma_m \right) x_{\text{zp}}^2$ at $n_p = 4.7 \times 10^6$.

In addition, the quadrature imprecision is below x_{zp}^2 at this point: $\langle \hat{X}_1^2 \rangle_{\text{imp}} = (0.57 \pm 0.09) \, x_{\text{zp}}^2$ (inset in Fig. 6.3c). This is approximately a factor of 200 above that of quantum-limited imprecision, which is consistent with the detection efficiency determined by $\kappa_R / \kappa \simeq 0.5$, the microwave loss between the device and amplifier ($\simeq 2$ dB), and the noise temperature of the amplifier at 4 K stage.

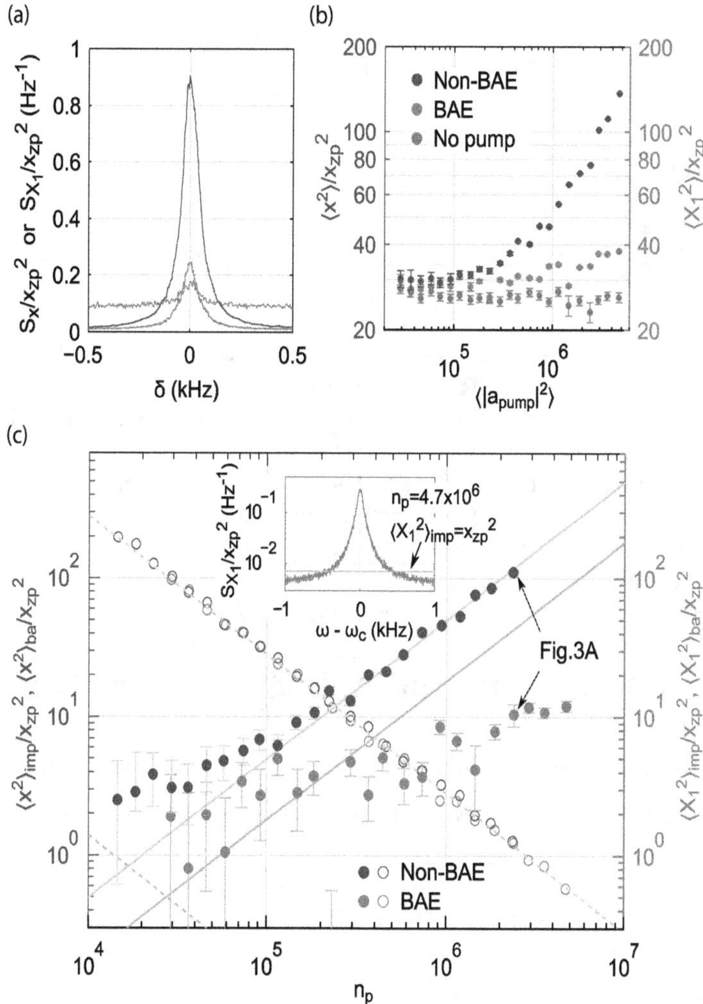

Figure 6.3: (a) The increase in the motional sideband area in DTT (blue) or in BAE (red), compared to the no balanced tones case (green), identifies the total measurement backaction. The spectra are centered at the mechanical resonance. (b) The mechanical fluctuations measured in DTT (blue) and BAE (red) at different pump strengths of the balanced tones. The green dots are the corresponding mechanical fluctuations measured without the balanced tones. (c) Measure imprecision (circles) and measurement backaction (dots) of position measurement in DTT configuration (blue) and mechanical quadrature measurement in BAE configuration (red). The solid blue line represents a fit to the measured backaction, including classical noise in the cavity. The solid green line is the expected quantum backaction from microwave shot noise. The dashed blue line shows a fit to the measured imprecision, and the dashed green line is the imprecision below the zero point.

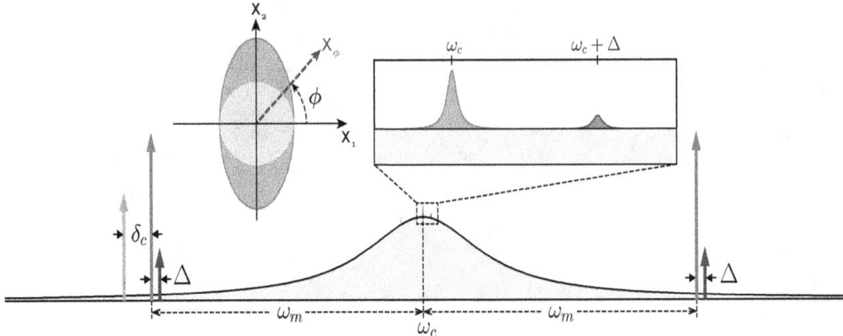

Figure 6.4: Schematic of the pump configuration in the double BAE experiment. The green arrow represents the cooling tone. The two red arrows represent the strong BAE pumps. The two blue arrows represent the weak BAE probe. The upper left inset is the schematic of the Wigner function, the green circle represents the thermal state cooled by the cooling tone (green arrow), the red ellipse represents the added noise from the BAE pump. (red arrows), and the blue dashed arrow represents the quadrature measured by the BAE probes. The upper right inset is the schematic of the mechanical sideband spectra. The red Lorentzian represents the sideband spectrum corresponding to the BAE pumps, and the blue Lorentzian represents the sideband spectrum corresponding to the BAE probes.

6.3 double BAE: Measurement of the quantum backaction

After demonstrating the backaction evading measurement of a single mechanical quadrature, in this section, we will probe the phase dependent backaction induced by the BAE measurement. As discussed in the previous section, the quadrature variances of the mechanical motion under a BAE measurement is given by

$$\langle \hat{X}_1^2 \rangle / x_{\text{zp}}^2 = 2n_m^{\text{th}} + 1, \tag{6.35}$$

$$\langle \hat{X}_2^2 \rangle / x_{\text{zp}}^2 = 2\left(n_m^{\text{th}} + n_{\text{ba}}\right) + 1, \tag{6.36}$$

where n_{ba} is the measurement backaction. The BAE measurement route all the backaction to the unmeasured mechanical quadrature $\hat{X}_2$. Therefore, the state of the mechanical oscillator under a BAE measurement is no longer a thermal state (the green circle in the upper left inset of Fig. 6.4). The BAE measurement breaks the rotational symmetry with the phase dependent measurement backaction. The mechanical quadrature variance at phase ϕ of this state is

$$\langle \hat{X}_\phi^2 \rangle / x_{\mathrm{zp}}^2 = \cos^2(\phi) \langle \hat{X}_1^2 \rangle / x_{\mathrm{zp}}^2 + \sin^2(\phi) \langle \hat{X}_2^2 \rangle / x_{\mathrm{zp}}^2$$

$$= \left(2n_m^{\mathrm{th}} + 1\right) + 2n_{\mathrm{ba}} \sin^2(\phi), \tag{6.37}$$

where $\hat{X}_\phi = \cos(\phi)\hat{X}_1 - \sin(\phi)\hat{X}_2$. The second term in Eq. (6.37) is the contribution of the phase dependent measurement backaction. In this section, we will detect this measurement backaction by applying an additional weak BAE measurement.

Fig. 6.4 is the schematic of the pump configuration. Similar to the BAE measurement, a red detuned tone (green arrow) is applied to damp the mechanical motion, and the green circle in the upper left inset represents the Wigner function of the mechanical state cooled by the cooling tone. In addition to the cooling tone, a pair of strong balanced tones (red arrows) is driving the cavity at $\omega_c \pm \omega_m$ to generate the phase dependent measurement backaction (second term in Eq. (6.37)). The red part of the ellipse in the upper left inset represents the measurement backaction. In order to detect the measurement backaction, another pair of weaker balanced tones (blue arrows) are driving the cavity at $\omega_c + \Delta \pm \omega_m$ with phases equal to $\pm\phi$ (relative to the strong BAE pumps (red arrows)) as a BAE probe to detect the mechanical quadrature $\hat{X}_\phi$ (the blue dashed arrow in the upper left inset). The detuning $\Delta \gg \Gamma_m$ such that there is no interference between the BAE pump and the BAE probe. The power of the BAE probe is kept sufficiently small such that the backaction induced by the BAE probe is negligible. The upper right inset is the schematic of the mechanical sidebands.

Fig. 6.5, a and b compare the measured quadrature variance of the BAE measurements from the BAE pumps (red circles) and the BAE probes (blue circles) at different phase ϕ. At $\phi = 0$, both the BAE pumps and BAE probes are measuring the minimum mechanical quadrature X_1, demonstrating the avoidance of the backaction from the noise in the detected microwave field. At $\phi = \pi/2$, the BAE probes are measuring the orthogonal mechanical quadrature X_2 while the BAE pumps are measuring the mechanical quadrature X_1. The maximal increases in fluctuations in the BAE probes signal demonstrate the measurement backaction from the BAE pumps. The blue curves in Fig. 6.5a(b) are the fits to Eq. (6.37)

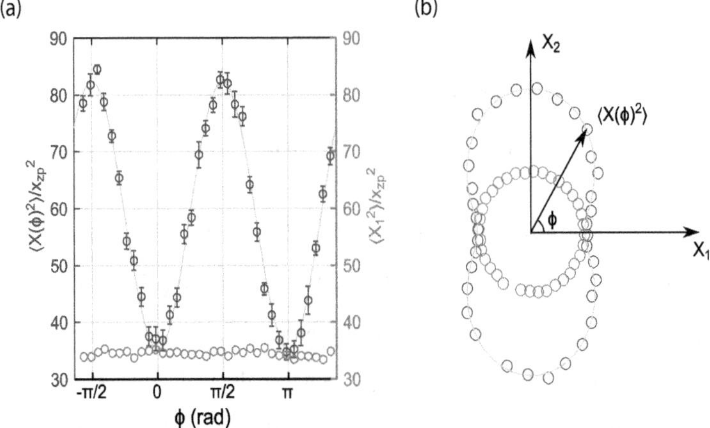

Figure 6.5: (a) An example of measured mechanical fluctuations along the BAE pump axis (red circles) and the probe axis (blue circles). ϕ is the angle between these two axes. The blue line is a fit to Eq. (6.37). (b) Polar plot of (a), defining X_1 and X_2 along the direction of minimum and maximum fluctuation, respectively.

with n_m^{th} and n_{ba} as fit parameters. The measurement backaction in the X_2 quadrature is

$$\langle \hat{X}_2^2 \rangle_{\text{ba}}/x_{\text{zp}}^2 = \langle \hat{X}_{\phi=\pi/2}^2 \rangle/x_{\text{zp}}^2 - \langle \hat{X}_{\phi=0}^2 \rangle/x_{\text{zp}}^2 = 2n_{\text{ba}}. \tag{6.38}$$

As discussed in section 2.9.2, the measurement backaction follows the relation

$$2n_{\text{ba}} = \frac{\Gamma_{\text{opt}}}{\Gamma_m} \left(2n_c^{\text{th}} + 1 \right), \tag{6.39}$$

where the term proportional to $2n_c^{\text{th}}$ is the classical backaction associated with the classical microwave noise and the term proportional to the $+1$ is the quantum backaction associated with the quantum fluctuations of the microwave field.

To explore the relation Eq. (6.39) of the measurement backaction, one can change the cavity occupation n_c^{th} by injecting microwave noise from room temperature into the device with the noise injection circuit described in the previous chapter. Fig. 6.6a shows the quadrature dependent noise at different injected noise power, the measurement backaction increases with the injected noise power. To measure the cavity occupation n_c^{th},

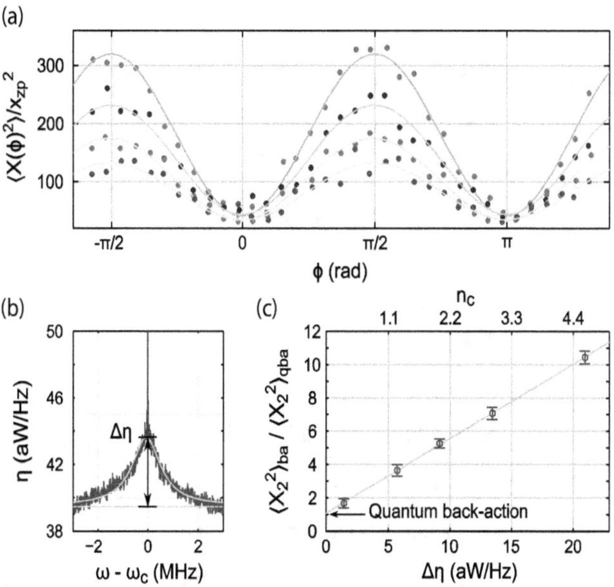

Figure 6.6: (a) Mechanical fluctuation along the probe axis at different microwave noise powers: $\Delta\eta$ = 5.71, 9.17, 13.41, and 20.99 (± 0.04) aW/Hz (brown, green, blue, and red dots, respectively). (b) Noise spectrum of the cavity resonance with occupation n_c^{th} = 0.9. $\Delta\eta$ is the noise density from the noise floor at the cavity resonance, which is proportional to n_c^{th}. The sharp peak at the center is the mechanical sideband at $n_p = 1.3 \times 10^6$. (c) Backaction in the X_2 quadrature normalized by the quantum backaction $\langle \hat{X}_2^2 \rangle_{\text{qba}}$ as a function of the noise density from the noise floor at the cavity resonance $\Delta\eta$.

we concurrently measure the output noise power density of the cavity resonance (Fig. 6.6b)

$$\bar{S}_{\text{out}}\left[\omega_c\right] = \mathcal{G}\left[\omega_c\right]\hbar\omega_c\left[\bar{S}_{\text{add}} + \frac{1}{2} + n_R^{\text{th}} + \frac{4\kappa_R}{\kappa}\left(n_c^{\text{th}} - n_R^{\text{th}}\right)\right],\tag{6.40}$$

and the system noise floor

$$\bar{S}_{\text{out}}^{\text{thru}}\left[\omega_c\right] = \mathcal{G}\left(\omega_c\right)\hbar\omega_c\left(\bar{S}_{\text{add}} + \frac{1}{2}\right)\tag{6.41}$$

by switching to the through port with the microwave switch. After subtracting off the noise floor, we obtain a linear relation between the cavity occupation and the measured noise density

$$n_c^{\text{th}} = \alpha\Delta\eta + \beta,\tag{6.42}$$

where $\Delta\eta = \bar{S}_{\text{out}}\left[\omega_c\right] - \bar{S}_{\text{out}}^{\text{thru}}\left[\omega_c\right]$, $\alpha = \kappa/4\kappa_R\mathcal{G}\left[\omega_c\right]\hbar\omega_c$ and $\beta = \left(\frac{4\kappa_R-\kappa}{\kappa}\right)n_R^{\text{th}}$. The measurement backaction is related to $\Delta\eta$ by

$$\frac{\langle X_2^2\rangle_{\text{ba}}}{\langle X_2^2\rangle_{\text{qba}}} = 2n_c^{\text{th}} + 1 = 2\alpha\Delta\eta + \left(1 + \beta\right),\tag{6.43}$$

where $\langle X_2^2\rangle_{\text{qba}} = \Gamma_{\text{opt}}/\Gamma_m$ is the quantum backaction. Fig. 6.6b shows the observed $\langle \hat{X}_2^2\rangle_{\text{ba}}$ versus $\Delta\eta$, which follows the expected linear relation. From the linear fit (blue line), we extract the slope $\alpha = 0.22 \pm 0.02\,(\text{aW/Hz})^{-1}$ and $\beta = 0.1 \pm 0.1$. The extrapolation to $\Delta\eta = 0$ demonstrates the measurement backaction induced by the quantum fluctuation from the microwave field. The nonzero β indicates that there is excess noise at the output port of the device.

To estimate the excess noise at the output port n_R^{th}, we measure the microwave noise spectrum without any microwave pump. In this setup, we assume the cavity occupation is solely due to the noise radiating into the device from the output port of the device, so that $n_c^{\text{th}} = \frac{\kappa_R}{\kappa}n_R^{\text{th}}$. This noise at the output port of the device generates a dip in the broadband noise floor (Fig. 6.7)

$$\bar{S}_{\text{out}}\left[\omega\right] = \frac{1}{\alpha}\left[\frac{\kappa^2}{\kappa^2 + 4\left(\omega - \omega_c\right)^2}\left(\frac{\kappa_R}{\kappa} - 1\right)n_R^{\text{th}} + \left(\frac{\kappa}{4\kappa_R}\right)\left(n_R^{\text{th}} + \frac{1}{2} + \bar{S}_{\text{add}}\right)\right].\tag{6.44}$$

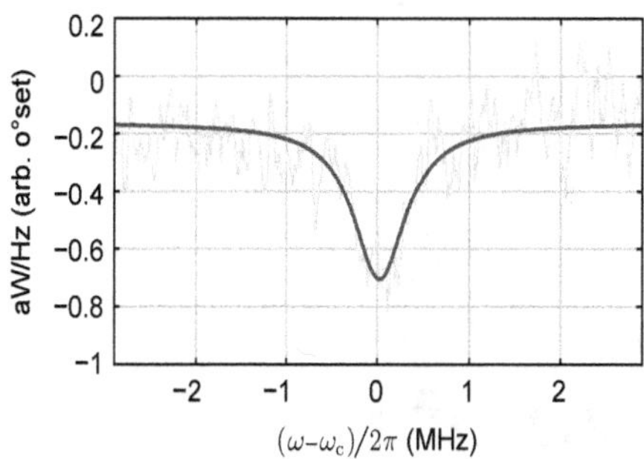

Figure 6.7: Cavity noise spectrum with no pump and no injected noise.

Taking $\kappa_R = 2\pi \times 450$ kHz from an independent measurement at 300 mK and fitting the measured spectrum with Eq. (6.44) (blue curve in Fig. 6.7), we extract $\kappa = 2\pi \times 860$ kHz and $n_R^{\mathrm{th}} \simeq 0.2$ (or $T_R = 50$ mK), which gives $\beta \simeq 0.2$, consistent to our observation 0.1 ± 0.1. This excess noise at the output port of the device maybe due to the power dissipated by the circulators at 100 mK stage.

Chapter 7

Mechanical squeezing

In this chapter, we will implement the reservoir engineering technique based on two-tone driving [ ] to generate and stabilize a quantum squeezed state of a micron-scale mechanical oscillator in an electromechanical system. In the first part of the experiment, we quantify the quadrature variances with the output spectrum of the electromechanical system and demonstrate the generation of the mechanical quantum squeezed state of a macroscopic mechanical object. In the second part of the experiment, we apply an independent backaction evading (BAE) measurement to directly quantify the squeezing. From the BAE measurement, we observe 4.7 ± 0.9 dB of squeezing below the zero-point level, surpassing the 3 dB limit of standard parametric squeezing techniques. The BAE measurement also reveals evidence for an additional mechanical parametric effect is observed. The interplay between this effect and the electromechanical interaction enhances the amount of squeezing obtained in the experiment.

As discussed at the end of the last chapter, an excess noise ($\sim$ 50 mK) at the output port of the device is observed, which maybe due to the power dissipated to the cryogenic circulators at the 100 mK stage. In order to remedy this issue, we move the circulators to the mixing plate in the following experiments (Fig. 3.6b). After relocating the circulators, the noise dip (Fig. 4.7) at the cavity resonance disappear. The cavity noise spectrum is flat in the absence of microwave drive and injected noise and we assume $n_R^{\text{th}} = 0$ throughout the experiments discussed in this chapter. The experiments are performed with device 2, which has the following parameters:

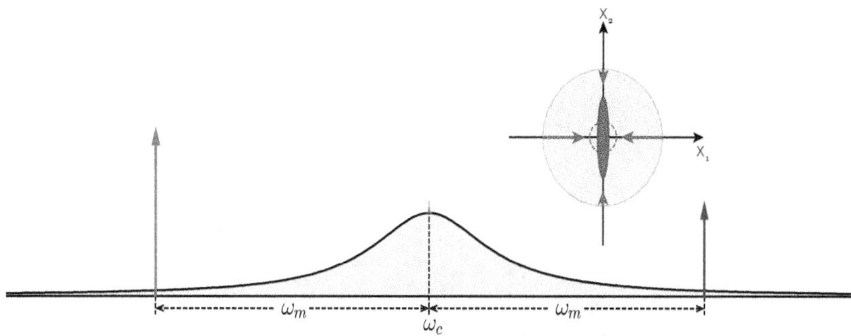

Figure 7.1: Pump schematic of the reservoir engineering configuration. The inset is the schematic of dissipative mechanical squeezing. The gray circle represents the initial thermal state in phase space. The engineered reservoir generates phase dependent dissipation that relaxes the mechanics into a squeezed state, which is represented by the blue ellipse. The gray dashed circle represents the zero-point level.

Parameter	Value	Descriptions
$\omega_m/2\pi$	5.8 MHz	Mechanical resonance frequency
$\Gamma_m/2\pi$	8 Hz	Mechanical linewidth
$\omega_c/2\pi$	6.08 GHz	Cavity resonance frequency
$\kappa/2\pi$	330 kHz	Cavity linewidth
$g_0/2\pi$	130 Hz	Electromechanical coupling
x_{zp}	~ 1.8 fm	zero-point motion

Table 7.1: Parameters of device 2.

7.1 Reservoir engineering mechanical squeezing

In this section, we apply the reservoir engineering technique introduced in section 2.9.3 to squeeze the mechanical motion. We drive the cavity with a stronger red detuned tone at $\omega_c - \omega_m$ (red arrow in Fig. 7.1) and a weaker blue detuned tone at $\omega_c + \omega_m$ (blue arrow in Fig. 7.1) to couple the cavity with the Bogoliubov mode of the mechanical oscillator. The cavity cools the Bolgoliubov mode with the dynamical backaction generated by the microwave drives and dissipatively drives the mechanics into a stationary squeezed

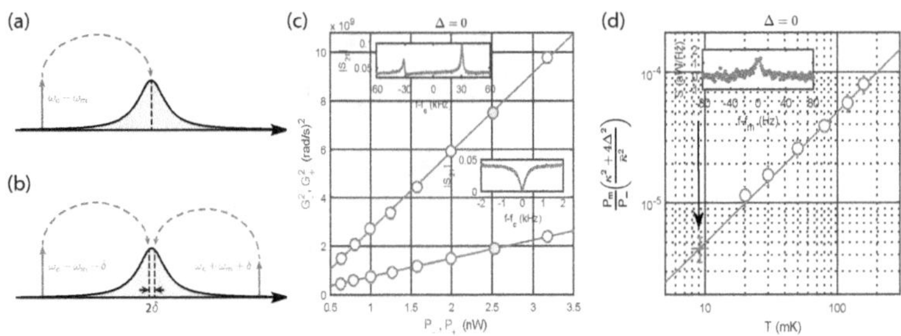

Figure 7.2: Calibrations of the mechanical squeezing experiment. (a) Pump configuration of the enhanced electromechanical coupling (G_-) calibration. (b) Pump configuration of the enhanced electromechanical coupling (G_+) calibration. (c) Calibrations of the enhanced electromechanical couplings $G_\pm$, the inserts are the transmission spectrums corresponding to the solid circles. (d) Calibration of the normalized motional sideband power, the insert is the sideband spectrum at the base temperature.

state (inset in Fig. 7.1). The mechanical sidebands of the two drive tones generate the Bolgoliubov mode signal at the center of the cavity resonance, which enables us to extract the quadrature variances of the mechanical squeezed state from the output noise spectrum of the electromechanical system.

7.1.1 Calibrations

In the experiment, we measure the output noise spectrum of the electromechanical system driven by the two drive tones

$$\bar{S}_{\text{out}}[\omega] = \mathcal{G}[\omega_c]\hbar\omega_c \left\{ \bar{S}_{\text{add}} + \frac{1}{2} + \kappa_R \bar{S}\left[G_-, G_+, \Delta, \delta, \kappa, \Gamma_m, n_c^{\text{th}}, n_m^{\text{th}}\right] \right\}, \qquad (7.1)$$

where $\mathcal{G}[\omega]$ and $\bar{S}_{\text{add}}$ are the total gain and the added noise at the output of the measurement chain, which are flat throughout the bandwidth of the experiment. $\bar{S}\left[G_-, G_+, \Delta, \delta, \kappa, \Gamma_m, n_c^{\text{th}}, n_m^{\text{th}}\right]$ is the noise spectrum of the electromechanical system which is given by Eq. (2.173). We spend an equal time interleaving measurement to measure the noise spectrum without any microwave drive to extract the noise floor

$$\bar{S}_{\text{out}}^0[\omega] = \mathcal{G}[\omega_c]\hbar\omega_c \left\{ \bar{S}_{\text{add}} + \frac{1}{2} \right\}. \qquad (7.2)$$

After subtracting the noise floor from the output noise spectrum, we obtain

$$\Delta \bar{S}_{\text{out}}[\omega] = \bar{S}_{\text{out}}[\omega] - \bar{S}^0_{\text{out}}[\omega] = \mathcal{G}[\omega_c]\,\kappa_R\hbar\omega_c\,\bar{S}\left[G_-, G_+, \Delta, \delta, \kappa, \Gamma_m, n_c^{\text{th}}, n_m^{\text{th}}\right]. \qquad (7.3)$$

In order to fit the spectrum $\Delta \bar{S}_{\text{out}}[\omega]$ to quantify the mechanical squeezing, an independent measurement of the prefactor $\mathcal{G}[\omega_c]\,\kappa_R\hbar\omega_c$ is required, which can be obtained from the calibrations of the enhanced electromechanical couplings and the thermal calibration.

7.1.2 Calibration of the enhanced electromechanical couplings

To obtain the calibrations of the enhanced electromechanical couplings $G_\pm$, we measure the transmitted power of the microwave drives

$$P_\pm = \mathcal{G}[\omega_\pm]\,\kappa_R\hbar\omega_\pm\,\lambda[\omega_\pm]\,n_p^\pm, \qquad (7.4)$$

where $\lambda[\omega_\pm]$ are the correction factors due to the parasitic channel [ ] and $n_p^\pm$ are the intracavity pump photon number induced by the red/blue detuned drives. The square of the enhanced electromechanical couplings are linear to the transmitted pump powers

$$G_\pm^2 = g_0^2 n_p^\pm = a_\pm \times P_\pm, \qquad (7.5)$$

with the calibration factors $a_\pm = \dfrac{1}{\mathcal{G}[\omega_\pm]\kappa_R\hbar\omega_\pm}\dfrac{g_0^2}{\lambda[\omega_\pm]}$.

We start with the calibration of the enhanced electromechanical coupling G_- induced by the red detuned tone. To do that, a single red detuned tone is applied at $\omega_c - \omega_m$ with transmitted power P_- (Fig. 7.2a). A network analyzer is used to generate a weak probe and sweep it through the center of the cavity resonance to measure the transmission spectrum of the mechanical sideband. The enhanced electromechanical coupling G_- can be extracted by fitting the transmission spectrum with the electromechanical model Eq. (2.171). By measuring the transmission spectrum with various transmitted power P_- and fitting with the linear relation (7.5) (the red line in Fig. 7.2c), we obtain the calibration $a_- = (7.49 \pm 0.10) \times 10^{17}\ \text{rad}^2\text{s}^{-1}\text{W}^{-1}$ and the intrinsic mechanical linewidth $\Gamma_m = 2\pi \times 8$

Hz.

A similar method is used to calibrate the enhanced electromechanical coupling G_+ induced by the blue detuned tone. In this case, a blue detuned tone is placed at $\omega_c + \omega_m + \delta$ with transmitted power P_+, where $\delta = 2\pi \times 30\text{kHz} \ll \kappa$. Since the blue detuned tone would amplify the mechanical motion and narrow the mechanical linewidth, the mechanical resonator becomes unstable when the cooperativity $C_+ = \frac{4G_+^2}{\kappa\Gamma_m}$ approaches to unity. In order to keep the mechanics stable, a constant red detuned tone is applied at $\omega_c - \omega_m - \delta$ to damp the mechanical motion (Fig. 7.2b). Similar to the calibration of G_-, we measure the transmission spectrum and extract the enhanced electromechanical coupling rate G_+ by fitting the transmission spectrum with the electromechanical model Eq. (2.171). By measuring the transmission spectrum at various transmitted power P_+ and fitting with the linear relation (7.5) (the blue line in Fig. 7.2c), we obtain the calibration $a_+ = (3.23 \pm 0.07) \times 10^{18}\,\text{rad}^2\text{s}^{-1}\text{W}^{-1}$.

7.1.3 Thermal calibration

Having calibrated the enhanced electromechanical couplings, we turn to the thermal calibration of the motional sideband noise power. To do that, a single red detuned tone is placed at $\omega_- = \omega_c - \omega_m$ with sufficiently small pump power such that the electromechanical damping effect is negligible ($\Gamma_{\text{opt}}^- = \frac{4G_-^2}{\kappa} \ll \Gamma_m$). We then measure the noise power of the up-converted motional sideband P_m, over a range of calibrated cryostat temperature T (Fig. 7.2d). Due to the weak temperature dependence of the cavity linewidth κ, we monitor the cavity linewidth at each measurement temperature. The resulting normalized sideband power is given by

$$\left(\frac{\kappa}{\bar{\kappa}}\right)^2 \frac{P_m}{P_-} = b_- \left(\frac{2}{\bar{\kappa}}\right)^2 \frac{k_B T}{\hbar\omega_m}, \tag{7.6}$$

where P_- is the transmitted power of the red detuned tone, $\bar{\kappa}$ is the average value of the cavity linewidth over the respective temperatures and $b_- = \frac{G[\omega_c]\omega_c}{G[\omega_-]\omega_-} \frac{g_0^2}{\lambda[\omega_-]}$ is the thermal calibration. The linear fit in Fig. 7.2d gives $b_- = (2.53 \pm 0.07) \times 10^5\ (\text{rad/s})^2$, which enables us to convert the normalized noise power into quanta. The prefactor $G[\omega_c]\kappa_R\hbar\omega_c$ is given by the ratio of the thermal calibration b_- and the calibration of the enhanced electromechanical coupling a_- (i.e. $\mathcal{G}[\omega_c]\kappa_R\hbar\omega_c = b_-/a_-$), which allow us to relate the measured noise spectrum and transmitted powers to the electromechanical model

$$\Delta\bar{S}_{\text{out}}[\omega] = \frac{b_-}{a_-}\bar{S}\left[G_-, G_+, \Delta, \delta, \kappa, \Gamma_m, n_c^{\text{th}}, n_m^{\text{th}}, \omega\right]. \tag{7.7}$$

7.1.4 Experimental results

7.1.4.1 Sideband cooling

Before squeezing the mechanical motion, a sideband cooling experiment is performed to characterize the device performance. To do that, a single red detuned drive is applied at $\omega_c - \omega_m$ at various pump power. The output spectrum of the mechanical sideband and the cavity resonance are measured to quantify the effective mechanical occupation $\bar{n}_m$. Fig. 7.3a shows the mechanical sideband spectra in the weak coupling regime. The optical damping rate increases with the intracavity pump photon number (n_p) and broadens the mechanical linewidth. As shown in Eq. (2.203), the power of the mechanical sideband is proportional to $\bar{n}_m - 2n_c^{\text{th}}$ due to the noise squashing effect. In order to extract the effective mechanical occupation, we concurrently measure the noise spectrum of the cavity resonance to obtain the cavity occupation n_c^{th}. Fig. 7.3c shows the effective mechanical occupation (red circles) and the cavity occupation (blue circles). As shown in Fig. 7.3c, the cavity occupation increases with the intracavity pump photon number, which limits the minimum achievable mechanical occupation. At $n_p = 4.1 \times 10^4$, the noise squashing $2n_c^{\text{th}}$ is larger than the mechanical occupation $\bar{n}_m$, which generates a noise dip at the cavity resonance (blue circles in Fig. 7.3a). As we further increase n_p, the noise squashing increases and finally evolves into normal mode splitting in strong coupling regime ($G \gg \kappa$), as shown in Fig. 7.3b. In strong coupling regime, the output noise spectrum is not simply a mechanical sideband sits on top of the center of the cavity noise. In this regime, we use the full formula Eq. (2.202) to fit the output spectrum and extract the phonon bath occupation n_m^{th} and the cavity occupation n_c^{th}, then calculate the effective mechanical occupation $\bar{n}_m^{\text{th}}$ with Eq. (2.196). As shown in Fig. 7.3c, the minimum mechanical occupation of this device is equal to 0.3 phonon, which is limited by the cavity heating effect. This cavity heating maybe due to

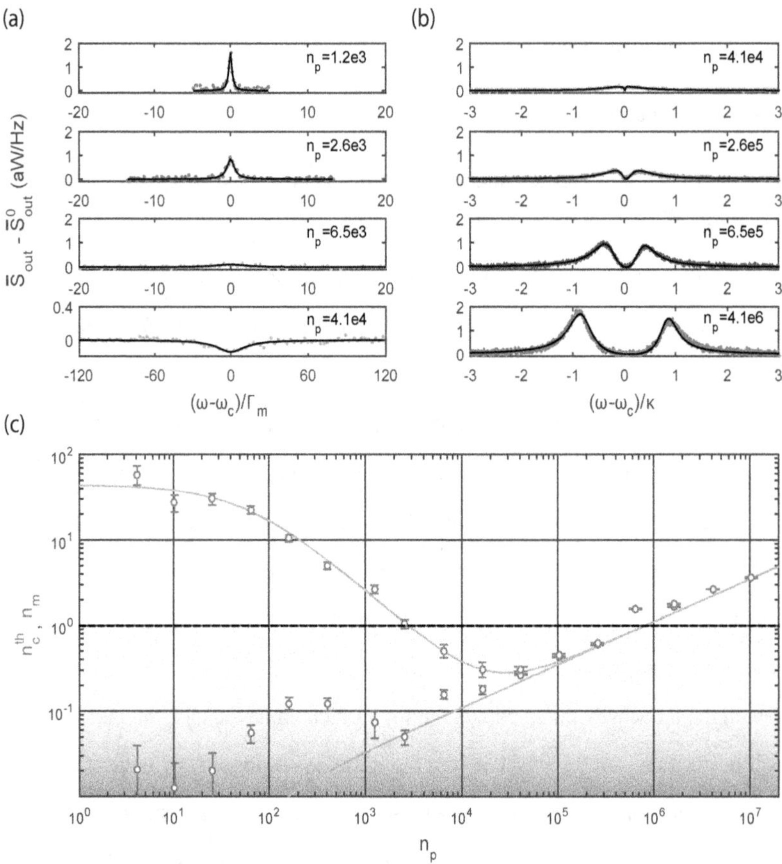

Figure 7.3: (a) Noise spectra of the up-converted mechanical sideband in weak coupling regime. (b) The noise spectra of the cavity resonance in the strong coupling regime. (c) Mechanical occupation (red squares) and cavity occupation (blue circles) extracted from the noise spectra.

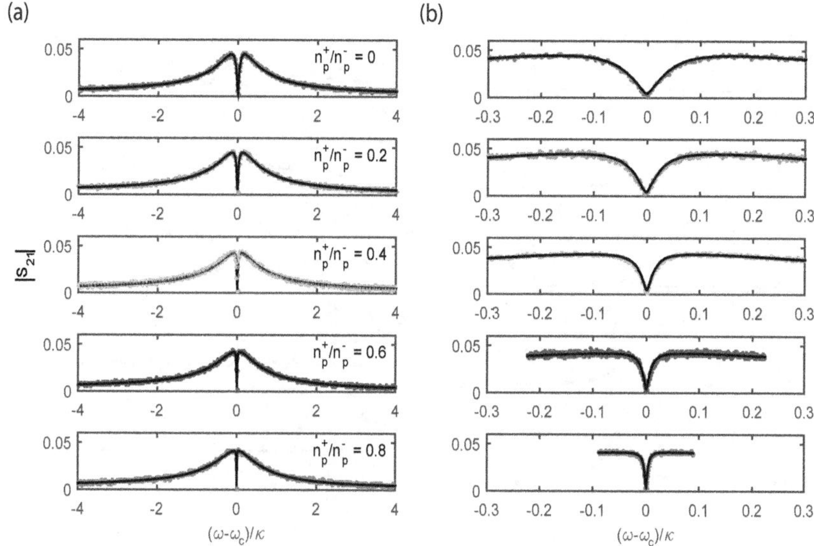

Figure 7.4: The driven responses at different pump photon ratios. (a) Driven responses in cavity span. (b) Driven responses near the interference signal.

the fluctuations of the two-level system (TLS) defects in the devices. A phenomenological TLS model predicts $n_c^{\mathrm{th}} \propto \sqrt{n_p}$ [ ], which is consistent with the observed cavity heating at large n_p (blue line in Fig. 7.3c).

7.1.4.2 Mechanical squeezing

Being able to cool the mechanical motion to the ground state is a good sign to generate the mechanical quantum squeezed state. To do that, we start ramping up the power of the blue detuned tone at $\omega_c - \omega_m$ to cool the Bolgoliubov mode of the mechanics and squeeze the mechanical motion. As discussed in section 2.9.3, the mechanical squeezing generated by two-tone reservoir engineering technique is very sensitive to the power ratio of the blue and the red detuned tones. In this experiment, we keep the total pump photon number $n_p^{\mathrm{tot}} = n_p^- + n_p^+ = 1.85 \times 10^5$, and squeeze the mechanical motion with various pump photon ratio n_p^+/n_p^-.

At each pump photon ratio, we measure the transmitted pump power of the microwave drives, from which we obtain the enhanced electromechanical couplings $G_\pm$ with the cal-

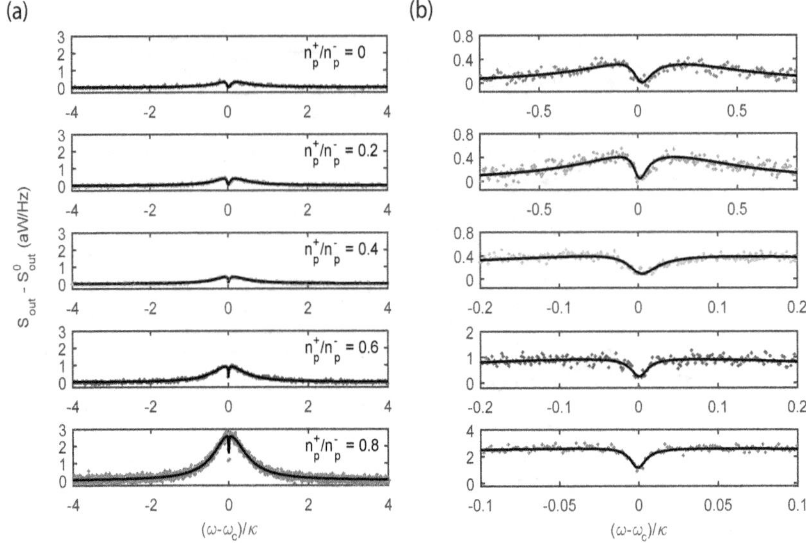

Figure 7.5: Output noise spectrum at different pump photon ratios. (a) Output noise spectra of the cavity resonance. (b) The noise spectrum near the mechanical sidebands.

ibrations $a_\pm$. The resonance frequencies of the cavity and the mechanical mode can be changed with the powers of the microwave drives due to thermal effects [ ] or nonlinearties [ ]. In order to precisely align the frequencies of the drive tones at $\omega_c \pm \omega_m$, we measure the driven response (S_{21}) of the system to extract the detunings and correct the frequencies of the drives. Fig. 7.4a shows examples of the measured driven responses (dots) and the corresponding fits (solid curves) at various pump photon ratios. The mechanical sidebands and the probe field interfere destructively at the center of the cavity resonance and generate a Lorentzian dip with linewidth equals to the effect mechanical damping rate $\Gamma_{eff} = \Gamma_m + 4G^2/\kappa$, which decreases as we increase the pump photon ratio. Fig. 7.4b shows the driven responses near the interference signals. By fitting the driven response data with Eq. (2.171), we can extract the detunings (δ, Δ), the mechanical resonance frequency (ω_m) and the frequency of the cavity resonance (ω_c), which enable us correct the frequencies of the drive tones. By iterating this procedures, we can precisely set the powers and the frequencies of the microwave drives.

Having determined the powers and frequencies of the two microwave drives, we measure

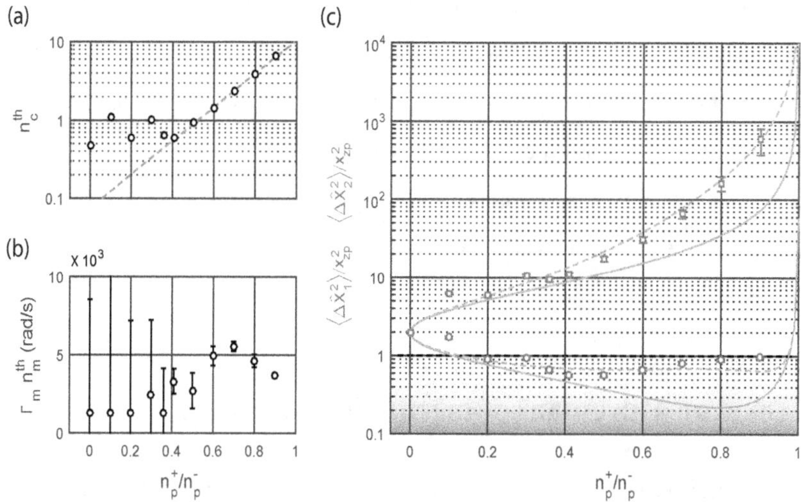

Figure 7.6: (a) Cavity occupation extracted from the noise spectrum. The dashed line is a linear fit of the pump ratio dependent heating. (b) Phonon bath heating rate extracted from the noise spectrum. (c) Squeezed quadrature variance (blue circles) and anti-squeezed quadrature variance (red squares) extracted from the noise spectra. The black dashed line indicates the quadrature variance at the zero-point level. The solid curves are the predictions from Eq. (2.241) with constant cavity and mechanical occupations extracted from the output spectrum at $n_p^+ = 0$. The dashed curves are the predictions including the cavity heating effect extracted from the experiment.

the output noise spectrum $\bar{S}_{\text{out}}$, interleaved with the unpumped noise floor $\bar{S}_{\text{out}}^0$. Fig. 7.5 shows examples of the measured output spectra with various pump photon ratios. As shown in Fig. 7.5a, the cavity heat up as we increase the pump photon ratio. Similar to the cooling spectrum (Fig. 7.3b), the noise squashing effect generates a Lorentzian dip with linewidth Γ_{tot} at the center of the cavity resonance (Fig. 7.5a). By fitting the spectra with the electromechanical model (2.243), we can extract the cavity occupation (Fig. 7.6a) and phonon bath heating rate (Fig. 7.6b).

With the extracted parameters from the driven responses and the output spectra, we can estimate the mechanical quadrature occupations with Eq. (2.238). Fig. 7.6c shows the quadrature occupations as a function of the pump photon ratio. The blue circles and red squares are the squeezed and anti-squeezed quadrature occupation respectively. The dashed line indicates the quadrature occupation at the zero-point level. The solid curves are the predictions from Eq. (2.241) with constant cavity and mechanical occupations extracted from the output spectrum at $n_p^+ = 0$; they agree with the data at low pump photon ratio. At large pump photon ratio, the cavity bath starts to heat up, which increases the mechanical quadrature variances. The dashed curves are the predictions including the cavity heating effect extracted from the experiment (dashed line in Fig. 7.6a). With the heating effect, the minimum quadrature variance is achieved at $n_p^+/n_p^- = 0.43$ with $\langle \Delta \hat{X}_1^2 \rangle = 0.56 \pm 0.02 x_{\mathrm{zp}}^2$, 2.5 ± 0.2 dB below the zero-point level.

7.2   BAE measurement of mechanical squeezing

While inferring the level of squeezing from the cavity output spectrum is convenient, it would be preferable to have a more direct method that does not rely on assumptions about the mechanical dynamics. This can be achieved in our system without needing to introduce an additional cavity resonance: we continue to use the cavity density of states near resonances to generate mechanical squeezing, but now use the density of states away from resonances to make an independent, backaction-evading measurement of a single mechanical quadrature. In this way, our single cavity effectively plays the role of two: it both generates squeezing, and permits an independent detection of the squeezing.

To directly detect the mechanical quadratures, in addition to the squeezing pumps, we introduce another pair of weak balanced probes (purple arrows in Fig. 7.7) at $\omega_c \mp \omega_m - \Delta$ with phase equal to $\pm\phi$ relative to the squeezing pumps (red and blue arrows in Fig. 7.7) to perform a backaction evading (BAE) measurement of the mechanical quadrature $\hat{X}_\phi$. In order to ensure no interference between the sidebands of the squeezing pumps and the BAE probes, we detune the BAE sidebands from the cavity resonance by $\Delta = 2\pi \times 160\mathrm{kHz} \gg \Gamma_{\mathrm{eff}}$. The power of the BAE probes are set about 10 dB weaker than the power of the squeezing pumps to avoid extra heating. In the experiment, the motional sideband spectrum of the BAE probes is measured (upper left inset in Fig. 7.7), from which we can extract the mechanical quadrature variance and linewidth. By sweeping the probe phase ϕ, we can perform tomography of the mechanical quantum state.

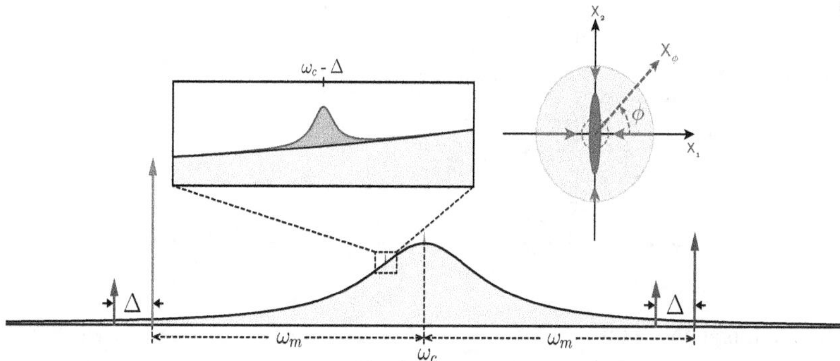

Figure 7.7: Pump schematic of the BAE measurement of mechanical squeezed state. The red and blue arrows represent the squeezing drives for the reservoir engineering technique. The two purple arrows represent the balanced drive for the BAE measurement. The upper left inset is the schematic of the BAE sideband noise spectrum. The upper right inset is the schematic of dissipative mechanical squeezing. The gray circle represents the initial thermal state in phase space. The engineered reservoir generates phase dependent dissipation that relaxes the mechanics into a squeezed state, which is represented by the blue ellipse. The gray dashed circle represents the zero-point level. The purple dashed arrow indicates the measured quadrature from the BAE measurement.

7.2.1   Calibrations

In this experiment, the detuning of the BAE sideband $\Delta = 2\pi \times 160\text{kHz}$ is comparable to the cavity linewidth, to precisely balance the BAE tones and correctly interpret the BAE noise spectrum, an independent calibrations of the enhanced electromechanical coupling rate $G_\pm$ and the normalized sideband power are necessary.

7.2.2   Calibration of the enhanced electromechanical couplings

We follow the same procedures in section 1.1.2 to calibrate the enhanced electromechanical couplings for the BAE measurement. To calibrate the enhanced electromechanical coupling G_-, we drive the system with a single red detuned tone at $\omega_c - \omega_m - \Delta$ with transmitted power P_- (Fig. 7.8a), and measure the driven response to extract the enhanced electromechanical coupling G_- (red circles in Fig. 7.8c) at various P_-. From the linear fit (red line in Fig. 7.8c), we obtain the calibration $a_-^{\text{BAE}} = (7.85 \pm 0.06) \times 10^{17}$. To calibrate the enhanced electromechanical coupling G_+, a constant red detuned tone is applied

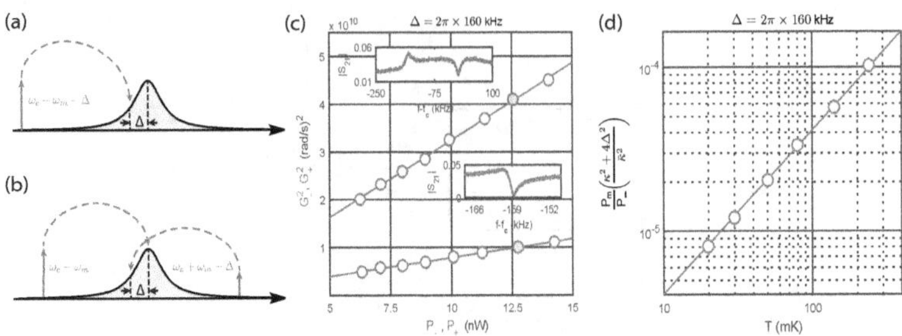

Figure 7.8: Calibrations of the backaction evading measurement. (a) Pump configuration of the enhanced electromechanical coupling (G_-) calibration. (b) Pump configuration of the enhanced electromechanical coupling (G_+) calibration. (c) Calibrations of the enhanced electromechanical couplings $G_\pm$, the inserts are the transmission spectrums corresponding to the solid circles. (d) Calibration of the normalized motional sideband power.

at $\omega_c - \omega_m$ to broaden the mechanical linewidth. Then we drive the system with a blue detuned tone at $\omega_c + \omega_m - \Delta$ with transmitted pump power P_+, and measure the driven response to extract the enhanced electromechanical coupling G_+ (blue circles in Fig. 7.8c) at various P_+. The linear fit (blue line in Fig. 7.8c) gives $a_+^{\mathrm{BAE}} = (3.24 \pm 0.03) \times 10^{18}$. The resulting calibrations $a_\pm$ enable us to precisely balance the BAE probes, i.e., $G_- = G_+$.

7.2.3 Thermal calibration

After obtaining the calibrations of the enhanced electromechanical couplings, a thermal calibration is used to covert the measured BAE sideband power to the mechanical quadrature variance. Similar to the thermal calibration in section 1.1.3, we drive the system with a single red detuned tone at $\omega_- = \omega_c - \omega_m - \Delta$ with sufficiently small pump power such that the electromechanical damping effect is negligible ($\Gamma_{\mathrm{opt}}^- = \frac{4G_-^2}{\kappa} \ll \Gamma_m$). We then measure the noise power of the up-converted motional sideband P_m, over a range of calibrated cryostat temperature T (Fig. 7.8d). The resulting normalized sideband power is given by

$$\left(\frac{\kappa^2 + 4\Delta^2}{\bar{\kappa}^2} \right) \frac{P_m}{P_-} = b_-^{\mathrm{BAE}} \left(\frac{2}{\bar{\kappa}} \right)^2 \frac{k_B T}{\hbar \omega_m}, \tag{7.8}$$

where P_- is the transmitted power of the red detuned tone, κ is the cavity linewidth, $\bar{\kappa}$ is the average value of the cavity linewidth over the respective temperatures and $b_- =$

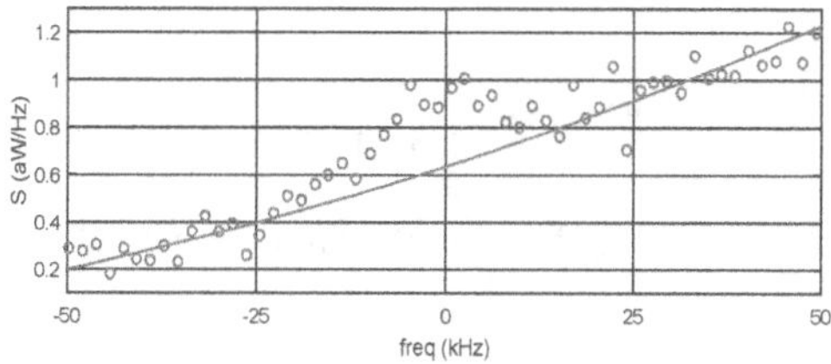

Figure 7.9: Example of the BAE noise spectrum. The red line is a background fit with a quadratic polynomial.

$\frac{\mathcal{G}[\omega_c - \Delta]\omega_c}{\mathcal{G}[\omega_-]\omega_-} \frac{g_0^2}{\lambda[\omega_-]}$ is the thermal calibration. The linear fit in Fig. 7.8d gives $b_-^{\mathrm{BAE}} = (2.77 \pm 0.04) \times 10^5 \ (\mathrm{rad/s})^2$.

Similar to the measurement of the squeezing output spectrum, we spend an equal time interleaving measurement to measure the pumped and unpumped noise spectrum. After subtracting the unpumped noise spectrum to remove the noise floor, the output noise spectrum of the BAE measurement is given by

$$\Delta \bar{S}_{\mathrm{meas}}^{\mathrm{BAE}} [\omega] = \bar{S}_{\mathrm{meas}}^{\mathrm{BAE}} [\omega] - \bar{S}_0 [\omega] = \bar{S}_c [\omega] + \bar{S}_{\mathrm{BAE}} [\omega], \tag{7.9}$$

the first term $\bar{S}_c [\omega]$ is the noise spectrum of the cavity resonance due to the non-zero cavity occupation. The second term $\bar{S}_{\mathrm{BAE}} [\omega]$ is the noise spectrum of the BAE sideband, which is given by

$$\bar{S}_{\mathrm{BAE}} [\omega] = \mathcal{G} [\omega_c - \Delta] \, \kappa_R \hbar \omega_c \frac{4 g_0^2}{\kappa} n_p \frac{\kappa}{\kappa^2 + 4\Delta^2} \frac{S_{X_\phi} [\omega]}{x_{zp}^2}, \tag{7.10}$$

where $S_{X_\phi} [\omega]$ is the mechanical quadrature spectrum. Because the BAE sideband is detuned from the cavity resonance with detuning comparable to the cavity linewidth, over the bandwidth of the BAE measurement, the cavity noise appears as a frequency dependent noise background. An example of the spectrum is given by Fig. 7.9, a quadratic polynomial is employed to fit the cavity noise background, as shown by the red curve in Fig. 7.9. The BAE sideband spectrum $\bar{S}_{\mathrm{BAE}} [\omega]$ is given by subtracting the noise spectrum from the fitted cavity noise background. The noise power of the BAE sideband is given by integrating Eq.

(7.10), which gives

$$P_m^{\text{BAE}} = \mathcal{G}\left[\omega_c\right] \kappa_R \hbar \omega_c n_p \frac{4g_0^2}{\kappa^2 + 4\Delta^2} \frac{\left\langle X_\phi^2 \right\rangle}{x_{zp}^2}.$$
(7.11)

With the thermal calibtration factor b_-^{BAE}, we can convert the normalized BAE sideband power to the quadrature variance

$$\frac{\left\langle X_\phi^2 \right\rangle}{x_{zp}^2} = \frac{1}{b_-^{\text{BAE}}} \left(\frac{4\Delta^2 + \kappa^2}{4} \right) \frac{P_m^{\text{BAE}}}{P_-}.$$
(7.12)

7.2.4   Experimental results

Fig. 7.10a is the mechanical quadrature variance as a function of the probe phase ϕ. We first perform the BAE measurement to a quantum squeezed state obtained with total pump photon number $n_p^{\text{tot}} = 1.85 \times 10^5$ and pump photon ratio $n_p^+/n_p^- = 0.41$ (the optimum squeezing point in Fig. 7.6c). The blue curve is the quadrature variance extracted from the output noise spectra (the yellow dots in Fig. 7.5). The blue circles are the quadratures variances obtained from the BAE measurement with probe photon number $n_p = 1.1 \times 10^4$. The minimum quadrature variance is achieved at $\phi = 0°$ with $\langle \Delta \hat{X}_\phi^2 \rangle = 0.34 \pm 0.07 x_{zp}^2$, 4.7 ± 0.9 dB below the zero-point level, which is lower than the quadrature variance inferred from the output spectrum, implying that there is additional dynamics at play (beyond the ideal electromechanical interaction). Then we perform the BAE measurement to a squeezed state obtained with smaller pump photon number ($n_p^{\text{tht}} = 1.35 \times 10^4$ and $n_p^+/n_p^+ = 0.5$). The red curve is the quadrature variance extracted from the output noise spectra and the red circles are the result of the BAE measurement with $n_p = 1.4 \times 10^3$. In this case, the results from the BAE measurement agree with the results obtained from the output spectra, which implies that the additional squeezing mechanism depend on the power of the intracavity fields.

The enhanced squeezing observed in the BAE measurement suggests an additional power dependent squeezing mechanism beyond the dissipative mechanism from electromechanical interaction; an obvious candidate is direct parametric driving of the mechanics. The presence of such driving is further corroborated by our observation of a phase dependence of

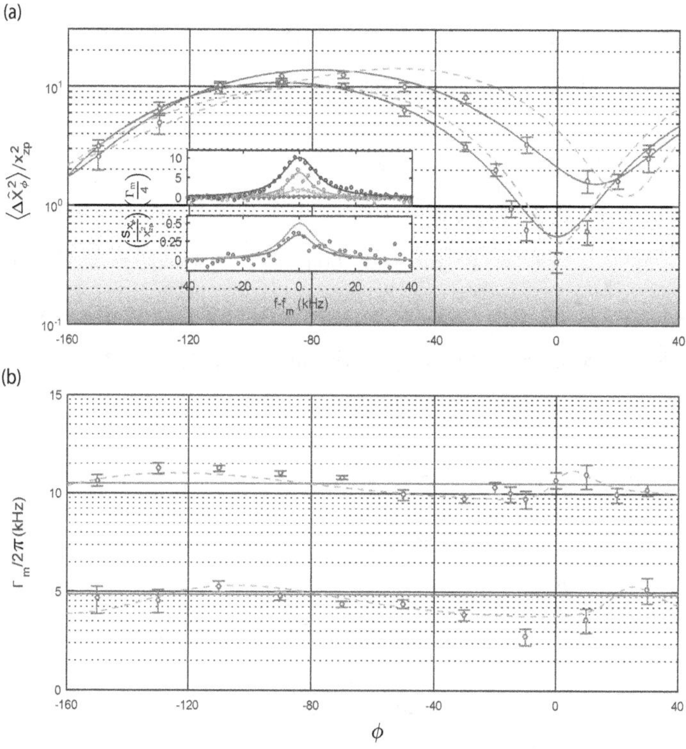

Figure 7.10: (a) Mechanical quadrature variance as a function of probe phase. The red (blue) circles are the quadrature variances of the weakly (strong) squeezed state as measured using the BAE technique. The solid curves are the quadrature variances inferred from the corresponding output spectra assuming no mechanical parametric drive. The dashed curves are the predictions of an electromechanical model including the mechanical parametric effect. The insets are the mechanical quadrature spectra of the strong squeezed state with phase ϕ at $-70°$ (red), $-50°$ (green), $-20°$ (yellow), $0°$ (blue). The gray Lorentzian in the lower inset represents the spectrum with quadrature variance equal to half of the zero-point fluctuation (the 3 dB limit). (b) Mechanical quadrature linewidth as a function of probe phase. The red (blue) circles are the measured mechanical quadrature linewidth of the weakly (strong) squeezed state. The solid lines are the theoretical predictions from the ideal electromechanical model. The dashed curves are the fit with the electromechanical model including the mechanical parametric interaction.

the quadrature linewidth in the BAE measurement (Fig. 7.10b). Similar induced mechanical parametric driving has been observed in other BAE measurements; they can arise via a number of mechanisms, including thermal effects as well as higher nonlinearities [,]. To understand the effects of this mechanical parametric driving, we phenomenologically add the parametric interaction to our otherwise ideal electromechanical model:

$$\hat{H}_{\mathrm{para}} = \hbar\lambda\left(e^{-i\psi}\hat{b}^2 + e^{i\psi}\hat{b}^{\dagger 2}\right), \tag{7.13}$$

where λ is the amplitude of the parametric interaction and ψ is the relative phase between the parametric drive and the squeezing pumps. Incorporating the parametric interaction into the ideal electromechanical Hamiltonian (3.19), the susceptibility matrix Eq. (3.28) becomes

$$\left(\chi\left[\omega\right]\right)^{-1} \equiv \begin{pmatrix} \chi_c^{-1}\left[\omega + \Delta\right] & 0 & -iG_- & -iG_+ \\ 0 & \chi_c^{-1}\left[\omega - \Delta\right] & iG_+^* & iG_-^* \\ -iG_-^* & -iG_+ & \chi_m^{-1}\left[\omega + \delta\right] & i2\lambda e^{i\psi} \\ iG_+^* & iG_- & -i2\lambda e^{-i\psi} & \chi_m^{-1}\left[\omega - \delta\right] \end{pmatrix}. \tag{7.14}$$

The solutions of the quantum Langevin equations are given by Eq. (3.29). With the solution of the phonon annihilation and creation operators, the mechanical quadrature $\hat{X}_\phi = \hat{b}e^{i\phi} + \hat{b}^\dagger e^{-i\phi}$ can be written as

$$\begin{aligned}
\hat{X}_\phi\left[\omega\right] =&\, D_{\phi,1}\left[\omega\right]\sqrt{\kappa}\hat{a}_{\mathrm{in}}\left[\omega\right] + D_{\phi,2}\left[\omega\right]\sqrt{\kappa}\hat{a}_{\mathrm{in}}^\dagger\left[\omega\right] \\
&+ D_{\phi,3}\left[\omega\right]\sqrt{\Gamma_m}\hat{b}_{\mathrm{in}}\left[\omega\right] + D_{\phi,4}\left[\omega\right]\sqrt{\Gamma_m}\hat{b}_{\mathrm{in}}^\dagger\left[\omega\right],
\end{aligned} \tag{7.15}$$

where $D_{\phi,j}\left[\omega\right] = \left(\chi\left[\omega\right]\right)_{3,j}e^{i\phi} + \left(\chi\left[\omega\right]\right)_{4,j}e^{-i\phi}$. The quadrature spectrum is given by

$$\begin{aligned}
\bar{S}_{X_\phi}\left[\omega\right] =&\, \frac{1}{2}\int\frac{d\omega'}{2\pi}\left\langle\left\{\hat{X}_\phi\left[\omega'\right],\hat{X}_\phi\left[\omega\right]\right\}\right\rangle \\
=&\, \left\{\left|D_{\phi,1}\left[\omega\right]\right|^2 + \left|D_{\phi,2}\left[\omega\right]\right|^2\right\}\Gamma_m\left(n_c^{\mathrm{th}} + \frac{1}{2}\right) + \left\{\left|D_{\phi,3}\left[\omega\right]\right|^2 + \left|D_{\phi,4}\left[\omega\right]\right|^2\right\}\kappa\left(n_m^{\mathrm{th}} + \frac{1}{2}\right),
\end{aligned} \tag{7.16}$$

which is a function of the enhanced electromechanical couplings (G_-, G_+), the detunings (δ, Δ), the thermal occupations (n_m^{th}, n_c^{th}), the amplitude and phase of the mechanical parametric drive (λ, ψ). The mechanical quadrature linewidth is given by fitting the predicted mechanical quadrature spectrum with a Lorentzian curve.

Having incorporated the mechanical parametric interation into the model, we can fit the observed phase-dependent quadrature linewidth with model to quantify the spurious parametric drive. Using the parameters (G_-, G_+, δ, Δ, n_m^{th}, n_c^{th}) extracted from the corresponding output noise spectrum, the quadrature linewidth can be written as a function of the probe phase ϕ, the amplitude λ and phase ψ of the parametric drive, i.e., $\Gamma_m\,(\phi, \lambda, \psi)$. By fitting the observed phase dependent quadrature linewidth data (Fig. 7.10b) with the function $\Gamma_m\,(\phi, \lambda, \psi)$, we can extract λ and ψ.

By assuming the phase of the parametric drive ψ follows the phase of the BAE probe (i.e. $\psi = \phi + \psi_0$, where ψ_0 is a constant phase shift), the model captures the observed phase dependence behavior of the quadrature linewidth, as shown by the dashed curves in Fig. 7.10b. From the fit, we extract $\lambda = 2\pi \times (121 \pm 34)$Hz, $\psi_0 = -121° \pm 52°$ for the weakly squeezed state (red dashed curve) and $\lambda = 2\pi \times (1.3 \pm 0.3)$kHz, $\psi_0 = -129° \pm 15°$ for the strong squeezed state (blue dashed curve). Surprisingly, if one instead assumes that the parametric driving is a result of the main squeezing tones (i.e. take ψ a constant independent of ϕ), one cannot capture the observed phase dependence of the quadrature linewidth. These results suggest that the parametric drive is induced by the BAE probes.

The dashed curves in Fig. 7.10a are the predicted quadrature variance including the mechanical parametric effect. The model suggests that the combination of the reservoir engineering with the mechanical parametric drive provide extra squeezing. However, the model doesn't fully capture the observed quadrature variance in the BAE measurement. The deviation may be due to the complicated heating effects associated with the underlaying nonlinearities [,] that cause the spurious mechanical parametric drive. We stress that our treatment of the spurious mechanical parametric drive is phenomenological; we do not know the precise microscopic mechanism which causes this driving. Nonetheless, it allows us to explain both surprising features of the BAE measurements (the observed phase-dependent mechanical quadrature linewidth, and the enhanced squeezing), and provide a direction to engineer the parametric drive to increase the squeezing.

In conclusion, we combine reservoir engineering and backaction evading measurement in a cavity electromechanical system to demonstrate a continuous QND measurement of mechanical squeezed states. A spurious mechanical parametric effect is observed and provide additional squeezing. Together with the spurious mechanical parametric drive, the reservoir engineering technique produce more than 3 dB squeezing below the zero-point level. The present scheme can be applied to generate and characterize more complicated quantum states by carefully engineering the nonlinear interaction [,]. The ability to generate and measure a strong quantum squeezed state of a macroscopic mechanical object would be useful for ultra-sensitive detection [ ], quantum information processing [ ], as well as the fundamental study of quantum decoherence [,].